Understanding
THE HUMAN BODY

The
Circulatory
System

Titles in the Understanding the Human Body series include:

Understanding
THE HUMAN BODY

The Circulatory System

Pam Walker and Elaine Wood

LUCENT BOOKS®

THOMSON
GALE

San Diego • Detroit • New York • San Francisco • Cleveland • New Haven, Conn. • Waterville, Maine • London • Munich

LIBRARY OF CONGRESS CATALOGING-IN-PUBLICATION DATA

Walker, Pam 1958–
 The circulatory system / by Pam Walker and Elaine Wood.
  p. cm. — (Understanding the human body)
 Summary: Discusses the organs and function of the human circulatory system, the vital functions of blood, and the medical diagnosis and treatment of heart disease and other circulatory disorders.
Includes bibliographical references.
 ISBN 1-59018-149-2
 1. Cardiovascular system—Juvenile literature. [1. Circulatory system.] I. Wood, Elaine, 1950–. II. Title. III. Series.
 QP103 .W354 2003
 612.1—dc21
                                                                2002009461

Printed in the United States of America

CONTENTS

FOREWORD

Since Earth first formed, countless creatures have come and gone. Dinosaurs and other types of land and sea animals all fell prey to climatic shifts, food shortages, and myriad other environmental factors. However, one species—human beings—survived the most recent millennia of evolution by adjusting to changes in climate and moving when food was scarce. The primary reason human beings were able to do this is that they possess a complex and adaptable brain and body.

The human body is comprised of organs, tissue, and bone that work independently and together to sustain life. Although it is both remarkable and unique, the human body shares features with other living organisms: the need to eat, breathe, and eliminate waste; the need to reproduce and eventually die.

Human beings, however, have many characteristics that other living creatures do not. The adaptable brain is responsible for these characteristics. Human beings, for example, have excellent memories; they can recall events that took place twenty, thirty, even fifty years earlier. Human beings also possess a high level of intelligence. Their unique capacity to invent, create, and innovate has led to discoveries and inventions such as vaccines, automobiles, and computers. And the human brain allows people to feel and respond to a variety of emotions. No other creature on Earth has such a broad range of abilities.

Although the human brain physically resembles a large, soft walnut, its capabilities seem limitless. The brain controls the body's movement, enabling humans to sprint, jog, walk, and crawl. It controls the body's internal functions, allowing people to breathe and maintain a heartbeat without effort. And it controls a person's creative talent, giving him or her the ability to write novels, paint masterpieces, or compose music.

Like a computer, the brain runs a network of body systems that keep human beings alive. The nervous system relays the brain's messages to the rest of the body. The respiratory system draws in life-sustaining oxygen and expels carbon dioxide waste. The circulatory system carries that oxygen to and from the body's vital organs. The reproductive system allows humans to continue their species and flourish as the dominant creatures on the planet. The digestive system takes in vital nutrients and converts them into the energy the body needs to grow. And the immune system protects the body from disease and foreign objects. When all of these systems work properly, the result is an intricate, extraordinary living machine.

Even when some of the systems are not working properly, the human body can often adapt. Healthy people have two kidneys, but, if necessary, they can live with just one. Doctors can remove a defective liver, heart, lung, or pancreas and replace it with a working one from another body. And a person blinded by an accident, disease, or birth defect can live a perfectly normal life by developing other senses to make up for the loss of sight.

The human body adapts to countless external factors as well. It sweats to cool off, adjusts the level of oxygen it needs at high altitudes, and derives nutritional value from a wide variety of foods, making do with what is available in a given region.

Only under tremendous duress does the human body cease to function. Extreme fluctuations in temperature, an invasion by hardy germs, or severe physical damage can halt normal bodily functions and cause death. Yet, even in such circumstances, the body continues to try to repair itself. The body of a diabetic, for example, will take in extra liquid and try to expel excess glucose through the urine. And a body exposed to extremely low temperatures will shiver in an effort to generate its own heat.

Lucent's Understanding the Human Body series explores different systems of the human body. Each volume describes the parts of a given body system and how they work both individually and collectively. Unique characteristics, malfunctions, and cutting edge medical procedures and technologies are also discussed. Photographs, diagrams, and glossaries enhance the text, and annotated bibliographies provide readers with opportunities for further discussion and research.

A Trek Through the Circulatory System

I

Compared to more complex creatures, a one-celled organism has a simple way of life. Materials that it needs for survival flow into the cell from the immediate surroundings. Similarly, all of the single-celled organism's wastes flow out into that same location. As long as a one-celled creature is floating in a favorable environment, it flourishes.

Life is more complicated for multicelled organisms. These more complex creatures have far fewer cells in direct contact with the environment. Therefore all cells do not have immediate access to life-sustaining oxygen and nutrients, nor can their wastes be disposed of in one simple step. As a result, multicelled organisms have developed an internal transport system that brings the external environment inside. This system delivers supplies to cells and simultaneously removes wastes. In humans and other animals this means of transport is the circulatory system.

The circulatory system has two primary jobs: It delivers life-sustaining materials to cells, and it removes wastes produced by those cells. Survival of cells in the body depends on a constant supply of oxygen and nutrients. Oxygen, a gas found in the atmosphere, and nutrients provided by food power cellular activities.

The circulatory system, which is also called the cardiovascular system, must be dependable. If any part of the circulatory system breaks down, body cells are deprived of life-sustaining oxygen and nutrients. The components of this distribution system that keep body cells energized include blood, a system of tubes that carry blood, and a pump so strong and efficient that it can propel blood in two separate circuits to nourish the whole body.

Blood is the medium in which the outside environment is carried deep within the body. The familiar thick, red liquid is made up primarily of red blood cells suspended in water. Red blood cells act like taxis on the blood highway as they pick up oxygen molecules in the lungs and deliver them to body cells. Waste gases are ushered out of the body by riding in blood on its return trip to the lungs.

The Road Well Traveled

Blood moves through the body in long, flexible hoses called blood vessels. Blood vessels carry blood from one end of the body to the other. If all the blood vessels in the human body were laid out end to end, they would extend more than sixty thousand miles, far enough to make two complete trips around the world.

Structurally, blood vessels are extremely durable tubes that surround an inner space or lumen. Rigid metal pipes could not withstand the stress that blood vessels must handle from before a person's birth until the moment of death. It is the unique design of vessels that provides much of their strength. The walls of most vessels have three distinct layers: inner, middle, and outer. A sheet of smooth cells forms the inner layer so that blood can flow through the vessels without hindrance. The middle layer, made of muscle and stretchy fibers, has the ability to expand (dilate) or narrow (constrict) to accommodate the flow of blood. The tough outer layer protects the vessel from injury and connects it to other tissues.

Blood vessels are classified into types: arteries, veins, and capillaries. Arteries have the crucial job of carrying blood away from the heart. They form the circulatory system's delivery route. Veins are the vessels that transport blood back to the heart. They serve as the return path. The body's largest arteries and veins originate in the heart and branch into smaller and smaller tubes. Big arteries divide into small branches called arterioles, and big veins subdivide into smaller venules. Capillaries, the smallest blood vessels, participate in this circuit to and from the heart on a micro level.

About seventy times a minute, fast-moving blood is propelled from the pumping heart into large arteries. To handle these strong surges of blood, arteries must be both tough and flexible. The layer of muscle in the walls of arteries is thick and easy to stretch, like a strong rubber band. With each spurt of blood, the arteries expand to absorb the energy, then recover by returning to their original size.

The muscular nature of the arterial walls contributes to the movement of blood through the body. With each beat, the heart forces blood out into large arteries, causing them to enlarge. When the heart relaxes, the arterial walls contract. These contractions squeeze the blood with enough force to push it farther along the vessels. If an artery is cut, blood shoots out in spurts caused by the rhythmic beating of the heart and expansion and contraction of the arterial vessel walls.

Arteries divide and subdivide, like branches on a tree. With each division the proportion of elastic fibers in their walls decreases, but the propor-

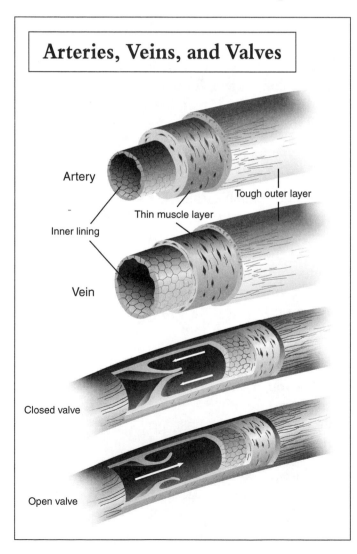

Arteries, Veins, and Valves

Artery

Tough outer layer

Thin muscle layer

Inner lining

Vein

Closed valve

Open valve

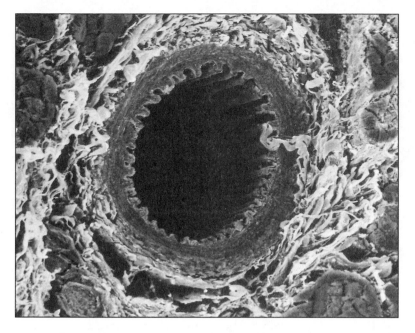

This cross section shows the dark center portion of the artery where blood flows.

tion of muscle increases. In the smallest divisions a single muscle may wrap around arterioles two or three times. Rhythmic squeezing and relaxing of these muscles help force blood into the smallest branches of the vessels, the capillaries.

Connecting Pathways

Capillaries are microscopic thin-walled vessels that measure only about one twenty-fifth of an inch thick. They are so small that ten of them lined up side by side would not be as thick as a hair. To pass through these tiny channels, red blood cells must line up single file. Despite their size, capillaries are the most important areas of the circulatory system. These tiny vessels are the sites where blood picks up or delivers materials. Tissues contain a dense network of capillaries so that no cell in the body is more than a millionth of an inch away from its blood supply. The entire circulatory mechanism works to keep enough blood in capillaries to meet the needs of cells.

Capillaries are only one layer thick, made entirely of smooth cells like those that form the inner lining of arteries

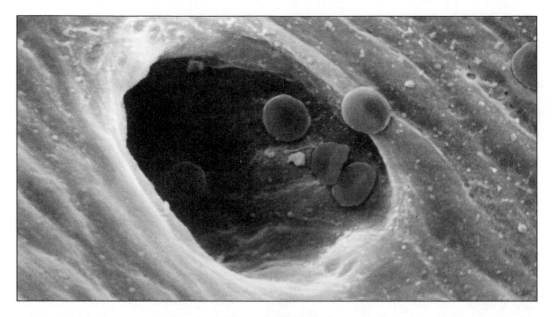

Tiny red blood cells travel into capillaries.

and veins. They deliver their cargo by maintaining a precise balance of fluid pressures between blood inside their walls and body fluid around the cells. Arterioles pump blood into capillaries with just enough force to drive plasma and dissolved nutrients through tiny vessels' thin walls. Cells, which are constantly breaking down food to generate energy, have a much lower concentration of oxygen than freshly oxygenated blood within the capillaries. It is a law of nature that gases always diffuse or flow from an area of high concentration to one of lower concentration. Therefore, oxygen in blood diffuses through the thin capillary walls into cells. At the same time, tissue fluid, which contains dissolved carbon dioxide and other wastes, takes the opposite route. It diffuses from cells, through capillary walls, into blood.

The Road Back Home

Blood traveling through capillaries flows into tiny venules which channel it into veins, the vessels that return it to the heart. Veins generally run parallel to arteries, creating two-way highways of north–south traffic throughout the body. Structurally, arteries and veins are similar, both having the same inner and outer layers.

By the time blood reaches veins, its flow has slowed dramatically, so blood in veins is pushed with less force than blood flowing through arteries. Consequently, veins are not under as much stress as arteries. As a result the layer of muscle in the walls of veins is thinner than the comparable layer in arteries. Walls of veins, with their load of slow-moving blood, do not have to be as strong and elastic as the walls of arteries. Additionally, the lumen of veins is larger than the lumen of arteries. Consequently, veins hold a larger supply of blood than arteries. About 65 percent of the body's blood is located within veins at any time.

Since veins do not expand and contract like arteries, they do not push blood along its way. The lack of a mechanism of propulsion can be a problem in areas where blood must flow against the pull of gravity, as it does when traveling from the feet back to the heart. However, venous blood does get some help. Movement of large muscles, such as those in the arms, legs, and abdomen, put pressure against veins and help keep blood moving. To prevent pooling of blood in lower regions of the body, veins contain valves. Valves are one-way doors that allow material to travel through, but prevent it from flowing backward. Valves are like footholds for blood on its return trip to the heart.

Blood does not follow a single continuous loop through the body. Because the heart works like two pumps running side by side, blood travels along two distinct circulatory paths. The pulmonary circulation carries blood from the right side of the heart to the lungs for oxygenation, then returns it to the left side of the heart. From there oxygenated blood is pumped along a route called the systemic circuit that extends to cells throughout the body. After a trip through the body, blood has lost its oxygen and is returned to the right side of the heart to repeat the cycle again.

A Short Trip

Once blood completes its route through the body, it has lost its oxygen supply and is carrying a heavy load of carbon dioxide and other wastes. About seventy times a minute, a shipment of "used" blood enters the right side of the heart, which pumps it directly to the lungs along the short pulmonary track. Within the lungs, arteries divide and subdivide into tiny capillaries whose sole job is to quickly absorb oxygen and release carbon dioxide before the next shipment of blood surges through.

The lungs are designed to make this high-speed exchange of gases possible. Oxygen-rich air enters the body through the mouth, travels down the windpipe, then enters the lungs through two tubes called bronchi. Bronchi divide and subdivide into millions of tiny branches called bronchioles, each of which ends in an air-filled sac called an alveolus. Capillaries wrap tightly around each tiny alveolus. Like capillaries, alveoli have very thin walls. Here, at the capillary and alveolus level, gases are exchanged.

This diagram shows the complex system of arteries that run through the body and into the heart, lung, and kidneys.

With each inhalation, the alveoli receive a fresh supply of oxygenated air. Because these sacs contain high levels of oxygen and the blood passing through the lungs is low in oxygen, oxygen diffuses from the air sacs of lungs into the capillaries. Blood in the capillaries, on the other hand, is carrying a concentrated shipment of carbon dioxide, but very little carbon dioxide is present in the air sacs. Consequently, this waste gas diffuses out of the blood into the alveoli. Carbon dioxide that accumulates in alveoli during this gas exchange is expelled from the lungs during exhalation. Reoxygenated blood travels back to the left side of the heart to prepare for distribution throughout the body.

The only time the pulmonary system is not used is during fetal development. Since a fetus does not breathe, there is no need for all of its blood to travel through the developing lungs. The fetus is supplied with oxygen through the placenta, a tissue that connects it to its mother's blood. Consequently blood returning to the right side of the fetal heart is directly shunted into the left side through a small opening, foramen ovale, to be included in the systemic circulation. At or soon after birth, this opening closes to provide efficient delivery of blood to newborn's lungs.

Around the Body

When oxygenated blood enters the left side of the heart from the lungs, it is ready to begin a much longer trip via the systemic circuit to cells in all parts of the body. The job of the systemic circulation is to deliver oxygen-rich blood throughout the body.

Along the way, blood acts as a shuttle service for products created by other body systems. For example, part of the systemic circuit includes the small intestine. This is where blood picks up the nutrients that it delivers to cells. As blood flows through a network of capillaries in the small intestine, molecules of nutrients from digested food travel through the intestinal walls and into capillary blood.

From the intestinal system, blood flows through the liver, where it is processed. The liver removes some nutrients, such as sugars, fats, and vitamins, and stores them until they are needed. It also cleans the blood by removing toxins and wastes produced by cellular activities.

Systemic circulation also functions as a pickup and delivery service for important cells and chemicals from other body systems. Hormones, chemical messengers produced by the endocrine system, are transported via the blood to specific destinations in the body where they help regulate body functions such as metabolism and growth. The immune system, which fights disease, also relies on the systemic circulation of the blood. It sends out specialized cells and proteins that safeguard the body against foreign invaders.

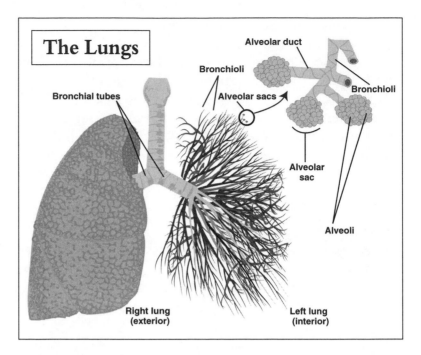

The Lungs

Bronchial tubes

Bronchioli

Alveolar sacs

Alveolar duct

Bronchioli

Alveolar
sac

Alveoli

Right lung
(exterior)

Left lung
(interior)

Along the systemic route, blood takes a trip through the kidneys. The two kidneys are bean-shaped organs located near the back of the abdominal cavity. They remove excess water and eliminate waste products that have accumulated in blood during the process of food digestion. The kidneys play a vital role in maintaining a proper balance of water, minerals, and nutrients within the body.

The Marvelous Pump

Circulation of blood through the body is possible because of a strong but delicate pump called the heart. The heart is a relatively simple, muscular pump whose job is to push blood through the vessels. Without the thrust provided by the heart's contractions, blood would pool in the lowest points of the body, unable to make the return trip to the lungs. This critically important organ is relatively small, measuring about the size of a man's clenched fist and weighing less than a pound. Yet muscle tissue in the heart works harder than any other muscle tissue in the body. Leg muscles tire quickly in a sprint, but even when the body is at rest, heart muscles work twice as hard as fast-pounding leg muscles. However, heart muscles never stop contracting, pumping relentlessly every minute of every day for seventy-five years or more.

The amount of blood that the heart pushes is phenomenal. When a person is resting, the heart pumps about two ounces of blood with every beat. This yields about five quarts of blood each minute, or seventy-five gallons of blood every hour.

Structurally, the heart looks a little like the standard shape seen in Valentine cards. Located between the lungs, it rests in the center of the chest cavity behind the sternum. It is situated so that its base, the apex, points toward the left side of the chest. Because the heart's rhythmic contractions can most easily be detected at the apex, many people wrongly assume that the heart is located on the left side of the chest.

A loose-fitting protective sac called the pericardium covers the heart. The outer layer of this covering, made of tough fibrous tissue, anchors the heart to the chest cavity. The delicate inner sac, which covers the heart itself, is actually a double membrane. The space between the two layers of the inner sac is filled with a lubricating fluid that prevents

A resin cast of the arterial blood supply to a kidney.

friction as the heart pumps in the chest cavity. It is the innermost layer of the double sac, the visceral pericardium, that actually covers the heart.

The walls of the heart are composed of three layers, very much like the structure of arteries and veins. The outermost layer of the heart wall, the epicardium, corresponds to the visceral pericardium. The thick middle layer is made of muscle

called myocardium. The thin innermost layer of the heart is smooth lining tissue, similar to the inner layer of vessels.

Cavities in the Heart

The interior space of the heart is divided into chambers. A partition called the septum divides the left side of the heart from the right. Blood from the two sides never mixes within the heart. Each half of the heart acts as an independent pump, serving different routes: The right half supplies the pulmonary circulation; the left, the systemic circulation. Each half is divided into a receiving chamber, or atrium, and a discharging chamber, or ventricle, giving the heart a total of four chambers.

Blood travels from the lungs to the heart's left atrium through four veins. The right atrium receives blood from three veins of the systemic circulation system one from the head region, one from the lower body, and one from the heart itself. Each atrium pumps the blood it receives into the ventricle directly below it.

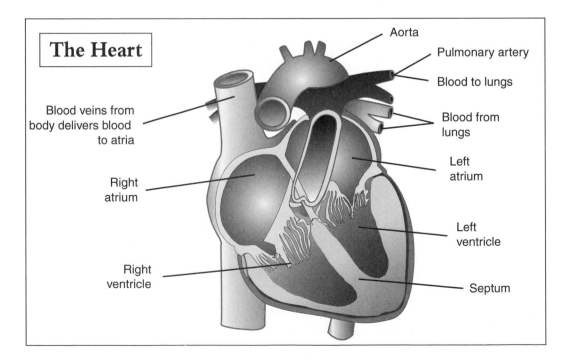

The Heart

Aorta

Pulmonary artery

Blood to lungs

Blood veins from body delivers blood to atria

Blood from lungs

Left atrium

Right atrium

Left ventricle

Right ventricle

Septum

Ventricles are larger and more muscular than atria because they have the job of propelling blood through arteries. Of the two, the left ventricle is the bigger. It sends blood along the systemic route to the far reaches of the body, a passageway that is longer than the path of blood through the lungs. When the right ventricle contracts, it forces blood into the arteries of the shorter pulmonary route that goes to the lungs.

Follow These Directions

When blood reaches the right atrium, it is returning from a trip through the body, so it has dropped off its supply of oxygen and is carrying a lot of carbon dioxide. The right atrium fills with blood, then contracts and forces the blood into the right ventricle. A strong contraction of the right ventricle then discharges blood into pulmonary arteries that carry it to the lungs.

The lungs provide a new supply of oxygen for the blood. Oxygenated blood travels back to the heart, this time filling the left atrium. A contraction of the atrium sends blood into the left ventricle. A ventricular contraction pushes blood into the aorta, the largest artery in the body, which branches into smaller and smaller arteries. Eventually blood enters capillaries, where it delivers oxygen to cells and picks up carbon dioxide. Deoxygenated blood, which is blood that is depleted of oxygen and is carrying a heavy load of carbon dioxide, travels back to the right atrium of the heart where it prepares for another trip to the lungs. Twenty seconds is all that is required for blood's complete trip through the body. Amazingly, all of the body's blood travels around the body three times each minute.

One-Way Lanes

The path of blood through the heart is a one-way trip. Valves in the heart play the same role as those in the legs, to ensure that blood does not flow backwards. Four valves are involved in maintaining the direction of blood flow: two atrioventricular valves and two semilunar valves.

An atrioventricular (AV) valve is located between each atrium–ventricle pair to prevent the backflow of blood each time the ventricles contract. The left atrioventricular valve, also called the bicuspid or mitral valve, has two flaps that close after blood has passed through it. The term "mitral" is often used to describe the bicuspid valve because it resembles the common two-flapped bishop's hat, the miter. The atrium and ventricle on the right side of the heart are separated by the right atrioventricular valve, also called the tricuspid because it is made of three flaps.

When the heart is relaxed (a period, or phase, called diastole), the two atrioventricular valves hang open, their flaps dangling down into the ventricles. During this relaxed time, blood flows into each atrium, then moves quickly into the ventricles. As ventricles contract in the phase called systole, they squeeze the blood and force it against the ventricle walls. This pressure pushes the valve flaps closed, shutting off the passages from the atria to the ventricles.

Each atrioventricular valve is held securely in place by strong cord–like tissue, sometimes referred to as the "heart strings." They connect the valves to the interior walls of the ventricles. These strings prevent the valves from collapsing back into the atria, like umbrellas turning inside out in a strong wind.

Blood leaving the heart from each ventricle passes through a set of semilunar valves, so named for their half-moon shapes. When ventricles contract, pressure forces blood out and pushes the semilunar valves open, flattening them back against artery walls. After blood surges out, the ventricles relax and valves shut, preventing any blood in arteries from streaming back into the heart.

Medical personnel can evaluate the health of a patient's heart by listening to its rhythm through a stethoscope. This instrument is made of hollow tubes that connect earpieces to a diaphragm. When placed on a patient's chest over the heart's apex, the diaphragm vibrates and creates the familiar "lub-dub" sounds of blood flowing through the valves. The first sound, "lub," occurs when blood hits the two AV (tricuspid and bicuspid) valves. The second, softer sound, "dub," is produced when the semilunar valves close.

An Electric Starter

Heart muscle, myocardium, is a unique tissue. It is incredibly strong, contracting or beating sixty to eighty times each minute of a person's life. Each contraction pushes blood through the body. For the heart to function correctly, its chambers must be stimulated to contract in a precise sequence.

The trigger that causes heart muscle to contract comes from the myocardium itself. A heart that has been surgically removed from a person's chest cavity will continue to beat. Heart muscle cells grown in the lab can be observed beating independently. This remarkable property is made possible by a built-in electrical conduction system that both starts and coordinates the beats. The conduction system originates in a group of cells called the pacemaker, or the sinoarterial (SA) node, located on the right atrium. The SA node is able to send an electrical impulse to both atria, causing them to contract at the same time. Contraction of the atria sends blood flowing through the open tricuspid and bicuspid valves into the ventricles.

By the time the ventricles have filled with blood, the stimulating electrical impulse reaches another specialized group of cells at the atrioventricular (AV) node, found near the middle of the heart between the atria and ventricles. The AV node then triggers each ventricle to contract, sending blood out of the heart into large arteries.

The Heart's Own Blood

Like all other cells in the body, the heart cells must be supplied with blood. However, blood traveling through the heart does not nourish heart tissue. The transport of blood to each heart cell is the responsibility of specialized vessels called coronary arteries.

Coronary arteries are so named because they encircle the heart like a crown. The left and right coronary arteries

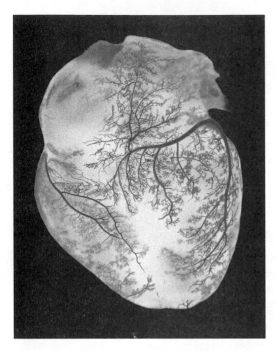

This view of a human heart shows both the arteries and veins.

branch off from the aorta. These two main arteries subdivide into smaller and smaller vessels, eventually forming thick capillary beds throughout the heart muscle. Blood leaves the heart tissue through small veins that empty into a large vein, the coronary sinus, which leads to the right atrium.

Just a Heartbeat Away

Three times a minute, blood is pumped through a double loop path in the body, traveling from the heart through the body, then from the heart to the lungs. The purpose of all this activity is to supply cells with oxygen and nutrients and to remove their wastes.

Blood is delivered to cells along a series of strong but flexible tubes called blood vessels. Large vessels subdivide into smaller and smaller branches until blood cells are forced to travel single file through the tiniest vessels in the body, the capillaries. It is at the capillary level that blood interacts with cells by delivering and picking up materials.

The pump that drives blood is powered by the body's strongest muscle. To gather fresh oxygen, the heart gently pumps half the body's blood volume to the lungs. Then it drives oxygenated blood through the systemic circuit with such force that, were the same energy used to send blood to the lungs, it would destroy them. Tirelessly, the heart keeps the red river of life flowing through all its vessels.

2 | Blood, the Red River

Throughout history, blood has been recognized as the river of life. Even the most primitive people knew that life ceased when blood stopped flowing through the body. Folklore is filled with stories of fictitious beings, such as vampires, that revered blood for its life-giving qualities.

Blood, a liquid connective tissue, is generously distributed all over the body. As a component of the circulatory system, blood carries out several vital roles. It helps maintain a stable environment for cells by providing them with everything they need to function, including oxygen and nutrients, and by carrying away their wastes. However, blood is more than just a supply line to cells; it has other important jobs, too. It contains cells of the immune system, the body's defense mechanism against disease and infection. Blood even carries the materials needed to stop a leak to prevent loss of this precious fluid.

The amount of blood in an individual varies somewhat with body size, but the average adult male contains about six quarts; an adult female, about five quarts. When a sample of blood is poured into a test tube and treated with salt, it separates into three distinct layers. The lightest part, made of pale yellow plasma, forms the top layer. Below the plasma is a thin, solid ring of white blood cells. On the bottom of the test tube is a thick band of red blood cells.

Proteins in the River

The liquid component, plasma, forms the stream through which blood's contents flow. Plasma is about 90 percent water. In a sample of blood, plasma makes up about 55 percent of the blood in volume, whereas cells make up about 45 percent. Dissolved within the plasma are a variety of substances, including proteins, nutrients, and electrolytes. The most abundant of these dissolved materials are the plasma proteins.

Plasma proteins are classified into three types: albumins, globulins, and fibrinogen. Albumins are small molecules, yet they make up about 60 percent of the dissolved proteins. Because they are so plentiful, albumins, along with the other plasma proteins, play a key role in maintaining the osmotic pressure of blood.

The movement of water into and out of blood vessels is controlled by a force called osmotic pressure. Water is the primary component of both blood plasma and a liquid called tissue fluid, which constantly bathes cells from outside the blood vessels. When osmotic pressure is the same inside and outside a vessel, the amount of water within the vessel stays the same. However, if the concentration of dissolved substances changes inside or outside of a blood vessel, osmotic pressure causes water to move. It travels through the cells of the vessels toward the region where the dissolved molecules are in highest concentration.

Therefore, it is very important that the levels of dissolved materials in blood, such as albumin, remain relatively constant. If they do not, water either leaves the vessels to move into tissues or leaves the tissues to move into vessels. When water leaves vessels, blood becomes thick and its flow is sluggish. On the other hand, loss of water from tissues can lead to dehydration. Globulins, which make up about 36 percent of plasma proteins, also play a role in balancing osmotic pressure. Globulins are described as three types: alpha, beta, and gamma. Made in the liver, alpha and beta globulins help transport molecules of fat and fat-like sub-

stances called cholesterol through the blood. Gamma globulins are made in tissues of the immune system. They
include the proteins that form the disease-fighting chemicals called antibodies.

The remaining 4 percent of plasma proteins is made up of
fibrinogen. Like alpha and beta globulins, this large molecule is produced in the liver. Fibrinogen is an essential
ingredient in the formation of blood clots. Blood clots play
roles as both good guys and bad guys. In their good role,
they prevent blood loss and aid in wound repair.
Unfortunately, they can also form inside of defective blood
vessels, interfering with the normal flow of blood.

Nutrients Catch a Ride

Proteins are not the only types of molecules dissolved in
plasma. This golden-colored liquid also holds nutrients
from digested food. Glucose, a simple sugar, is transported in plasma from the small intestine to individual cells
for use as fuel. When the body has an oversupply of glucose, the blood delivers some of it to the liver for storage
and the rest to cells that convert it to fat. Amino acids, the
building blocks of proteins, are also carried to the liver in
plasma. The liver may use them to make new proteins, or
break them down to supply energy when glucose or fat is
not available.

Plasma also contains lipids, compounds that are greasy to
the touch. The types of plasma lipids are triglycerides (fats),
phospholipids, and sterols, such as cholesterol. Since lipids
are not soluble in water, they are carried in plasma in special
protein packages called lipoproteins. Lipids play several roles
in the body. They are a rich source of energy and serve as a
chief component of cell membranes.

All of the cells found within blood originate in bone marrow, the fibrous tissue within the central cavity of bones
throughout most of the body. Red blood cells, white blood
cells, and platelets are derived from a single type of precursor cell, a hematopoietic stem cell. These originator cells are
incredibly fruitful; in less than a month, ten of them can

A red blood cell (left), a white blood cell (bottom center), and four platelets (right) all come from hematopoietic stem cells.

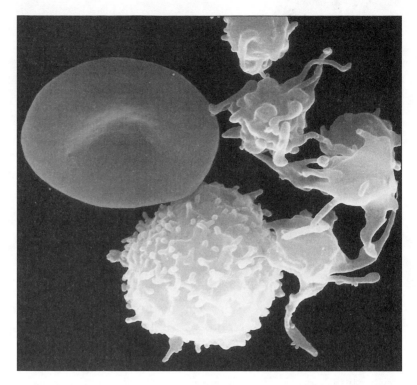

multiply into 30 trillion red blood cells, 30 billion white blood cells, and 1.2 trillion platelets, enough to replace every blood cell in the body.

Little Cells with a Big Job

Red blood cells are by far the most numerous of the solid ingredients of blood. One drop of blood contains more than 250 million red blood cells. Amazingly, red blood cells are some of the smallest cells in the body; a stack of five hundred would only measure .04 inch. However, the importance of these cells is reflected in their number instead of their size. About 25 trillion red blood cells can be found coursing through the body's vessels at any time.

The job of red blood cells, or erythrocytes, is to bind with oxygen so this essential gas can be delivered to body cells. These cells are uniquely specialized to fulfill this role. Each cell is a disc that is thick around the rim and thin in the center. This unusual configuration offers more surface area than a flat disc. The greater surface area enhances the red blood cells' ability to absorb and carry gases.

In addition, unlike most cells, mature red blood cells do not have a nucleus. Just before reaching maturity, these cells extrude their nuclei through their membranes. With the loss of nuclei comes an improvement in the red blood cells' ability to carry oxygen. For in the space previously occupied by the nucleus, each red blood cell can now hold hemoglobin, an oxygen-binding protein whose pigment makes the red cells red.

Hemoglobin, the Oxygen Magnet

Each red blood cell is composed primarily of water and hemoglobin. Hemoglobin molecules are very long, being made of more than ten thousand atoms. These atoms form four strands of amino acids called globins, each of which surrounds a disc of iron-containing heme.

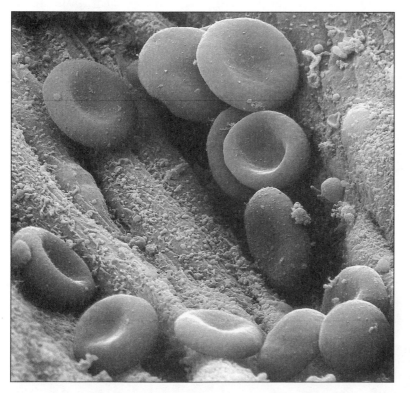

Red blood cells like these, are packed with hemoglobin.

When oxygen is plentiful, as it is when red blood cells travel through the capillaries in freshly oxygenated lungs, heme scoffs up oxygen molecules. When oxygen levels are low, heme releases its oxygen, allowing it to diffuse to areas like active muscles that need it most. If oxygen is not available at all, heme binds to carbon dioxide, a pairing that helps the blood carry this waste back to the lungs.

Without hemoglobin, blood's oxygen-carrying capacity would be dramatically reduced. Each of the four heme discs in a hemoglobin molecule can bind to an oxygen molecule. Every red blood cell contains 270 million hemoglobin molecules. Within an adult there are so many red blood cells that if they were stacked one on top of the other, they would reach thirty-one thousand miles into the sky.

The Life of a Red Blood Cell

Red blood cells live longer than most other types of cells, about 120 days after being released from the bone marrow. Each cell travels through the body at least three hundred thousand times before it dies. As red blood cells wear out, they are filtered from circulation by the spleen, which has this task as one of its specialties, and by the liver. Important components of red blood cells, such as iron, are recycled.

After a red blood cell dies, its hemoglobin is first separated into two parts, the iron-containing heme molecule and the globin protein. Heme is further broken down into iron and a green pigment called biliverdin. Some iron is stored in the liver for future use. The rest is recycled to the bone marrow for incorporation into new red blood cells. Biliverdin is converted into bilirubin, a red pigment that is excreted from the liver in bile.

More than 100 million red blood cells are made every minute, and an equal number are destroyed in that same amount of time. Thus the number of red blood cells circulating through the body remains quite stable. Even so, under some circumstances the body will alter the rate at which red blood cells are made. For example, if blood is lost due to injury or through blood donation, the kidneys release a hormone called erythropoietin, which stimulates marrow to speed up its production schedule. With erythropoietin on the job, red blood cell production may increase sevenfold.

The body uses the same strategy at high altitudes. In the thin air of mountaintops and other high altitude locations, there is not as much oxygen available to tissues as there is at sea level. The drop in oxygen levels within tissues triggers the release of erythropoietin, which travels in the blood to bone marrow. There it stimulates red blood cell production. Within a few days, new red blood cells appear in circulation. Eventually, the increased number of red blood cells boost the amount of oxygen in tissues, and production of erythropoietin stops.

The Mission of White Blood Cells

White blood cells, or leukocytes, are not as numerous as red blood cells. They number only one for every seven hundred red cells and most have life spans ranging from twelve hours to a few days. Their job is not related to carrying oxygen to cells; they protect the body against disease. These large cells lead very active lives pursuing and destroying invaders such as bacteria, viruses, protozoa, and fungi.

Whereas there is only one type of red blood cell, white blood cells exist in several forms. Granulocytes are large white blood cells containing vesicles that have a grainy appearance. Granulocytes patrol the body, searching for invading bacteria or viruses. If they encounter any, they surround and destroy the invaders. A granulocyte on patrol exhibits the sliding motion of an amoeba.

Another type of leukocytes contains no granular vesicles, and thus is called agranular. White blood cells of the agranular variety have much longer life spans than granulocytes, living for several months, and in some cases for years. Two types of agranular cells are monocytes and lymphocytes. Monocytes attack and ingest foreign cells, in much the same way granulocytes do. Lymphocytes act less directly: They produce antibodies, proteins that attack foreign substances that enter the body.

Plugging the Flow

Blood is a precious liquid, and the body cannot afford to lose much of it. Moderate blood loss can lead to illness, and extreme loss can result in death. To protect against blood deficit, the body has developed some strategies to staunch an unwanted flow.

Platelets, or thrombocytes, are the smallest cells in blood. Platelets are more common than white blood cells, but not as numerous as red. Formed in the marrow like red blood cells, platelets are produced from huge precursor cells called megakaryocytes. When these big cells mature, they shatter into thousands of platelike pieces with life spans of only five to eight days. These platelets are cell fragments rather than whole cells, but they have the ability to move through tissue in the same sliding fashion used by some white blood cells.

Platelets are involved in two strategies for stopping blood loss. When a blood vessel is damaged, platelets traveling in the blood quickly move to the site of injury. They adhere to the rough edges of the vessel within one to five seconds. Once stuck in place, platelets swell

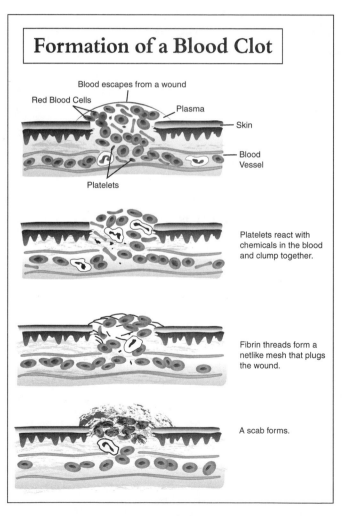

Formation of a Blood Clot

Blood escapes from a wound
Red Blood Cells
Plasma
Skin
Blood Vessel
Platelets

Platelets react with chemicals in the blood and clump together.

Fibrin threads form a netlike mesh that plugs the wound.

A scab forms.

and shoot out tiny spikes, forming a mass that stops small leaks.

If the tear in a vessel is too large for platelets alone to heal and bleeding continues, then Plan B goes into action. The damaged tissue releases a chemical cry for help in the form of a group of proteins called blood platelet factors. These chemicals set off a series of chemical reactions, which cause blood to coagulate or clot.

Prothrombin, one of the alpha globulins normally found dissolved in plasma, is one of the first responders. This molecule is converted to another form called thrombin. In turn, thrombin reacts with fibrinogen molecules, another one of the dissolved plasma proteins. This causes individual fibrinogen molecules to hook together, end to end, forming long threads. Fibrin threads stick to the damaged part of the vessel, creating a tangled net of fibers. As blood flows through the mesh, it captures other blood cells and platelets. The resulting mat forms a blood clot that stops the flow of blood.

The blood vessels themselves are also able to help slow the loss of blood. When a blood vessel is damaged, muscle making up the wall of that vessel reacts by contracting. This provides the immediate benefit of slowing the loss of blood. This initial damage-inspired contraction only lasts a minute or two. However, as platelets form a plug, they release a chemical called serotonin, which stimulates the muscle to continue contracting.

Normally, blood clots do not form inside of vessels. The smooth interior surface of the vessels, and the movement of the blood through them, does not allow thrombin levels to build up. That is why blood clots generally form in places where blood is flowing slowly, or not at all. However, damage inside a vessel can cause an increase in thrombin levels and lead to the formation of a clot. For example, when fatty deposits called plaque accumulate inside vessels, platelets may rush to the sites, causing the whole cascade of clotting reactions to begin again.

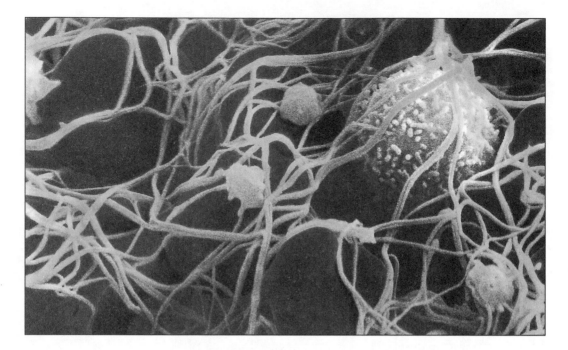

A weblike structure traps red blood cells preventing blood flow. The resulting blood clot can lead to illness and even death.

Just the Type

Even though early physicians knew that blood is the body's life-giving liquid, they did not understand the full array of its functions. In the Middle Ages, European physicians had little or no opportunity to learn about blood or any other part of the human body because the church forbade the dissection of corpses. However, by 1628 the ban on studying anatomy had been lifted. As a result, the English physician William Harvey was able to study the bodies of dead humans and animals. By doing so he figured out the path that blood takes as it circulates. Harvey, a pioneer in the study of blood, was one of the first scientists to explain the role of blood in the body. Shortly after Harvey's contributions, the first transfer of blood from one individual to another was attempted. This early transfusion of blood was not successful.

In spite of failures, scientists and physicians did not give up on the idea of blood transfusions. Since very few patients survived substantial loss of blood, transfusions seemed like the only hope. Therefore physicians continued their experiments to perfect the technique. In 1667 Jean-Baptiste Denis in France and Richard Lower in England attempted to transfuse blood from lambs to humans. Transfusions from animals to humans caused so many deaths that the practice was soon banned. In other research, milk was substituted for animal blood, but these transfusions also failed. Eventually, the practice of transfusion was prohibited.

In 1900 Karl Landsteiner, an Austrian physician, moved to the United States to conduct research on blood transfusions. Through his microscope Landsteiner was able to see different types of molecules on surfaces of red blood cells. He simply named one kind of molecule "A" and the other type "B." Landsteiner's discovery proved to be a life-saving breakthrough that made transfusions of blood possible. His work won him a Nobel Prize in medicine.

Today scientists know that the molecules identified by Landsteiner are chemical markers, which serve as identity tags. Different kinds of red blood cells have different markers. Landsteiner realized that a patient could only receive blood from someone whose red blood cells had the same chemical markers as their own cells.

English physician William Harvey.

Based on these markers, blood is classified into different types. In the study of blood the identifying surface molecules on red blood cells are called antigens. An antigen is any substance that stimulates the body to make antibodies, defensive proteins, against it. When antibodies attack the antigens on red blood cells, clotting of blood can occur. This can lead to serious health problems or death.

Landsteiner's work led to the classification of blood into types A, B, AB, and O. If red blood cells have antigen A on their cell membranes, the blood is called type A. Type B blood has antigen B on its red blood cell membranes. Some people have both types A and B antigens on their red blood cells, and their blood type is described as AB. On the other hand, there are others who do not have any antigens on their red blood cells, and they are said to have type O blood.

The Riddle Solved

Although the failure rates of early transfusions were tragically high, Landsteiner's research finally made it safe to give the blood of one person to another. Landsteiner found that antigens on donor red blood cells can react with antibodies that are already present in a recipient's blood. By the age of six months, a baby has naturally developed antibodies against the type of antigens she or he lacks.

The blood plasma of a person with antigen A does not contain antibodies to A; if it did, that person's blood would clot. However, the plasma of a person with antigen A does contain antibodies to antigen B. In other words, blood containing antigen A carries antibodies that are designed to destroy cells carrying antigen B. This explains why so many early transfusions failed.

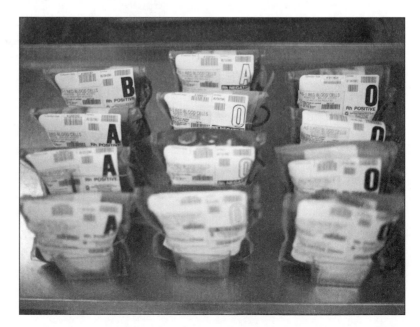

Bags of blood are labeled to show the blood type inside.

Likewise, the plasma of a person with type B blood contains antibodies to antigen A but not to antigen B. The blood plasma of people with type AB blood does not contain any antibodies. On the other hand, the blood plasma of people with type O blood carries antibodies to both A and B antigens.

Before a patient is given a blood transfusion, health care personnel test the patient's blood to determine its type. Then, knowing what kinds of antigens are on the patient's blood cells, they can order blood that has the same antigens.

For example, a person with type A blood can receive type A blood in a transfusion. Giving that person type B or AB blood would cause clotting, sickness, and possibly death. People with type A blood can be given type O blood, however, because it does not contain any antigens that can stimulate a clotting reaction. By the same token, a person with type B blood can receive type B blood or type O blood, but either type A or AB would lead to dangerous clotting.

People with type AB blood have an advantage because they can receive any type of blood. For this reason folks with type AB blood are often called universal recipients. People with type O blood can only receive type O blood. However, type O people can donate blood to anyone, and are known as universal donors.

The Monkey Contribution

After the identification of A, B, AB, and O blood types, tests were developed that allowed most people to receive blood transfusions without any problems. But physicians were surprised to find that some patients still had adverse reactions to blood transfusions. In 1939, research with rhesus monkeys led to a discovery that explained why: the existence of another blood antigen, which is called the Rh or D type.

The Rh antigen is present on red blood cells of 85 percent of the people in the world. Those who have the antigen are described as Rh+ (Rh positive) and those without it as Rh- (Rh negative). Whereas blood normally contains antibodies to antigens A or B, people with Rh- blood do not naturally have the antibodies to Rh factor. However, they will develop the antibodies if the Rh antigen is introduced to their bloodstream. Therefore blood transfusions from an Rh+ donor to a recipient who has antibodies to the Rh factor can damage the recipient's blood, thereby posing a serious danger to the person's health.

There are two situations in which the Rh factor can cause life-threatening complications in a developing fetus or newborn baby. In one scenario, an Rh- woman whose blood contains antibodies to the Rh factor conceives a child who has Rh+ blood inherited from the father. The mother may have developed these antibodies after exposure to the Rh antigen through a blood transfusion. These antibodies can travel from the mother to her developing child. If the Rh antibodies encounter the Rh+ fetal blood, the fetal blood can be damaged or destroyed. Unlike A, B, AB, and O antibodies, Rh antibodies do not trigger clotting. Instead, they can cause red blood cells to burst. In response, the infant's body speeds up its production of red blood cells, releasing them before they are mature and fully functional.

In another picture, an Rh- mother carrying a Rh+ baby may be exposed to a small amount of the baby's blood through tears in the placenta during delivery. Such exposure would stimulate production of antibodies against the Rh factor in the mother's blood. These antibodies would not hurt the mother because she does not have the Rh factor in her blood. They would not arise quickly enough to harm the fetus. However, problems could develop if the mother were to have a second child with Rh+ blood. In the second pregnancy the mother's blood would already contain antibodies to the Rh antigen. If any of her blood crossed the placenta to the baby, its antibodies could damage the baby's blood. To prevent such a tragedy, after delivering an Rh+ child, an Rh- mother is given an injection of anti-Rh vaccine. This halts Rh antibody production in her body.

The Gift of Life

Because loss of large amounts of blood can cause death, blood transfusions are an important part of medical treatment. Blood transfusions are critical for people injured in accidents as well as those undergoing surgery or receiving treatment for cancer or blood diseases. There is no substitute for human blood; it cannot be manufactured nor can it be replaced by using animal blood. A ready supply of human blood is critical to meet the needs of patients during medical emergencies.

According to the National Blood Data Resource Center, about 8 million blood donors supply enough blood for about 4.5 million patients each year. On an average, thirty-four thousand units of blood are needed in the United States each day. However, only about 5 percent of healthy Americans who are eligible to donate blood actually do so. To be eligible, most states require that donors be at least seventeen years old, although some states will accept donations from younger people with parental consent.

Some conditions make one ineligible to donate blood. People who have used illegal intravenous drugs cannot

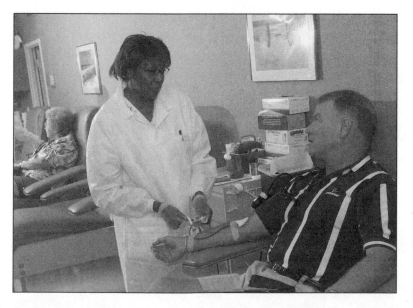

A man donates blood. Donated blood can be used in transfusions, which can save peoples' lives.

donate because they are at risk of carrying a blood-borne disease. Hemophiliacs, people who have a condition that prevents their blood from clotting normally, cannot give blood. Anyone who has had hepatitis, a serious liver condition, is also ineligible.

Blood is usually donated at community or hospital-based centers, or at schools and workplaces during blood drives. Bloodmobiles, specially modified buses that travel to malls, office complexes, and houses of worship, make donating convenient.

At the donation site, prospective donors are given educational material that contains information about the risks of blood donation. At this point prospective donors can leave without donating if they wish. A person who decides to stay is asked a series of questions to establish his or her health history. The next step is a short physical exam that checks blood pressure, pulse, and temperature. A few drops of blood are taken from a finger to determine whether or not the potential donor is anemic.

If a prospective donor passes all of these steps, he or she is asked to lie down or sit in a comfortable reclining chair. The skin of the inner arm is cleaned, then a medical specialist called a phlebotomist removes a new needle from its sterile wrapping and inserts it into a vein. The needle is connected to a rubber tube that leads to a blood bag. Typically, one unit, which is about one pint of blood, is collected. Afterward the donor is given refreshments and allowed a brief period of rest.

After blood is collected it is tested to determine its blood group and Rh type. It is then screened for the presence of several diseases including hepatitis B and C, Human Immunodeficiency Virus (HIV), and syphilis. Freshly collected blood is described as whole blood because it contains plasma, red cells, white cells, and platelets. Since many transfusion patients do not need all parts of the blood, it is common to separate blood into its different components. This process allows one unit of blood to be used to help several people.

The Life Blood

The red river that streams through veins, arteries, and capillaries is designed to support cells. Chock full of nutrients, oxygen, proteins, white blood cells, red blood cells, and platelets, blood reaches every cell of the body to supply its needs.

Red blood cells are distinctively designed to manage oxygen transport. Their unique disc shape provides them with maximum surface area for the easy diffusion of gases into and out of cells. And their interiors are spacious because they have shed their nuclei to provide room for extra hemoglobin, the protein that attracts and clasps oxygen.

Before 1900 scientists were not very successful in transferring blood from one patient to another. However, in modern surgery, replacement or supplemental blood is an essential supply. Blood is available to patients today because of the work of earlier researchers who discovered that there are different types of blood, identified by the antigens located on the red cells. Millions of blood donors generously share their gift of life with those who need it.

A scientist examines packets of blood plasma.

The Circulatory System's Inside Story

3

The circulatory system has provided more imaginative figures of speech than all of the other body systems. People may be described as "heartbroken" when they are sad, as "bighearted" when they are kind, or as "cold-blooded" if they are cruel. A scary event is "heart stopping" or "bloodcurdling," but a good story is "heartwarming." In reality, the blood and heart have nothing to do with how individuals feel. Emotions do not stem from the circulatory system; they originate in the nervous system. Yet a sad person is never described as "brain broken" or "heavy-brained."

Because of this supposed link of the heart to emotions, many people have some false ideas about the circulatory system and how it works. Understanding the biology behind confusing "symptoms" can help clear up misconceptions.

The Heart Throb

The heart may be associated with feelings because strong emotions can cause it to flutter or skip a beat. The degree to which people are sensitive to their own heartbeat varies tremendously from one individual to another. Some are acutely aware of their own heart rhythm while others rarely notice it. There are certain times when almost everyone feels their heartbeat, such as during strenuous exercise. Most folks can sense their heart beating simply by lying on their left side.

Occasionally, one's heart rhythm changes from a regular pulsing to an irregular thumping. People who have the erroneous idea that a brief change in beat indicates danger may become alarmed. However, premature heartbeats occur in perfectly healthy people. These temporary adjustments of heartbeat are normal. Any such change or adjustment in the heart's rate of contraction is called a palpitation or arrhythmia.

There are several forms of arrhythmia. Normally, the adult heart beats about 60 to 100 times per minute. A very fast heart rate, more than 130 times per minute, is called tachycardia. On the other hand, a slow heartbeat is described as bradycardia. Although changes in rhythm because of outside influences such as exercise, pain, or excitement are normal, prolonged bouts of arrhythmia may actually indicate heart disease and warrant attention. Thus, sensations of abnormal heart rate do not imply that emotions or feelings originate in the heart. Instead, the heart, like other body systems, responds to impulses generated by the brain as it deals with life issues.

The Blue Bloods

Another long held but false idea about the circulatory system has to do with blood. At one time it was thought that different classes of people had different colors of blood. Many believed that noblemen and other people with elevated status in society had blue blood rather than red, the color of blood in common people. Consequently, members of high society were referred to as the "blue bloods." Although speculations about the sociocultural reasons for this description can be interesting, the term itself is not scientifically accurate.

Blood is never blue. However, the color observed in the veins of light-skinned people may have led to the term "blue blood." Blood vessels, when stripped out of the body for scientific study, are actually shiny and white. Thus the red blood flowing through arteries and veins is darker than the vessels themselves and changes their appearance to the observer. Additionally, light that strikes blood vessels is diffused or scattered by the skin, contributing to the illusion that the blood is a blue color, very much for the same reason that the sky appears to be blue but actually is not.

Blue Is Not Always a Good Thing

The body works hard to maintain a constant internal temperature. Important chemical reactions cannot occur in the body if its temperature gets too high or too low. Body temperature is determined by how much heat is lost and how much is produced in the body.

The skin and tissues are kept at a constant temperature by the flow of blood through them. Blood is warmed by the breakdown of food inside cells. Any increase in body temperature triggers a rise in the amount of blood flowing to the skin. Additionally, the rate of flow accelerates, which helps to return body temperature to normal. This boost in blood circulation enables the body to lower its core temperature by radiating heat outward.

Extremely low temperatures can prevent crucial internal chemical reactions from occurring.

When the skin is exposed to very cold conditions, the body loses heat. In response, vessels in the skin are constricted and blood flow is diverted to the brain and other vital organs. However, this protection of the core temperature comes at a price: Extremities like fingers and toes cool very rapidly. Consequently, they can easily be damaged by frigid temperatures.

Hypothermia, an abnormally low body temperature, can cause the circulatory and other body systems to slow. Severe hypothermia can lead to heart attack and death. Symptoms begin with cold skin, shivering, and drowsiness, but can escalate to confusion, seizures, stupor, and coma. Hypothermia can be prevented by wearing layered clothing, eating ample amounts of food, and drinking a lot of liquids. Contrary to some popular opinions, alcohol and caffeine should be avoided in cold weather. They dilate vessels, causing blood to move away from the body core to skin and extremities, cooling the body.

Although blood itself is not blue, the lips or fingers of a person who is very cold may develop a blue coloration. Cyanosis, a bluish tinge of the skin and mucous membranes, occurs when the blood contains abnormally high levels of deoxyhemoglobin, which is hemoglobin that has lost its oxygen. In very cold conditions a slowed circulatory system is not the only adjustment the body can make to conserve heat. Superficial blood vessels constrict causing blood flow to slow. In turn, blood travels through the lungs less rapidly than normal. At the same time, cells remove extra oxygen from blood in an effort to generate more internal heat.

The skin also appears blue if blood supply is restricted. If blood cannot reach tissues, cells in that area use up all of the oxygen available in hemoglobin, turning the bright red protein the characteristic darker shade of deoxyhemoglobin. Blue color in tissue is a warning sign that oxygen levels are low. Sometimes objects can restrict blood flow. For example, a rubber band wrapped tightly around a finger can prevent oxygen from reaching the end of the finger. Unless the band is removed quickly, there can be tissue death at the end of the finger.

Big and Little Blue Veins

The venous system is made up of a network of small blood vessels that merge together to form larger ones. Like tributaries of a river, these vessels eventually create a flow that takes blood from the far points of the body back to the heart. Blood in the venous system is often moving against gravity.

Since the walls of veins are not muscular and strong like the walls of arteries, veins cannot constrict to help push blood along its way. Yet, even though muscle movements in the legs, abdomen, and chest press against veins and help move blood in the direction of the heart, and one-way valves keep blood from flowing backward, veins do not always move blood efficiently. Some veins are weak and expand when filled with blood. Others contain faulty valves that cannot hold blood in place. In either case veins can expand and form swollen areas that look like thick cords under the skin. These areas are referred to as varicose veins. They usually occur in the large, superficial veins of the legs.

The causes of varicose veins are poorly understood. At one time it was believed that this condition resulted from long periods of standing. Today doctors know that this is not the case. Obesity, pregnancy, and constriction of veins by wearing tight clothes may contribute to varicose veins. However, the deciding factor appears to be a tendency to inherit the problem.

Varicose veins may cause mild pain or an itchy feeling, but they rarely require treatment. Medical intervention may be needed if they interfere with circulation enough to produce excessive swelling. The most common remedy is surgical removal of the veins. An alternative to surgery is injection with a salt solution or some other agent that causes the veins in that area to dry up and shrink. Since most varicose veins lie just under the skin and not deep within muscle,

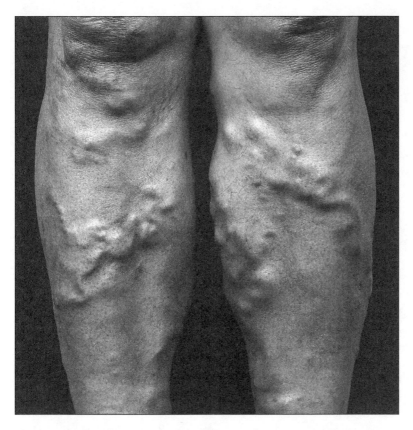

Varicose veins are believed to be inherited.

they are not the primary carriers of venous blood. Thus, blood has plenty of alternate paths to the heart.

Spider veins are often found in people who have varicose veins. Spider veins are small vessels just under the skin that look like long, blue spidery legs. Once believed to be made up of burst capillaries, doctors now know that the causes of spider veins are more complex. Since they appear more often in women than in men, their cause may be related to female hormones. If these veins cause itching or pain, or if they are unattractive, they also can be treated by saline injections.

Another Color Indicates Problems

Yellow is another color that can indicate a problem somewhere in the body. Physiologic jaundice is a condition in which the skin and the whites of the eyes become yellow.

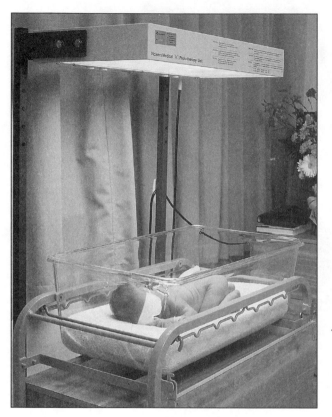

Bright light is used to dissipate bilirubin, which causes neonatal jaundice.

It is caused by an accumulation of bilirubin, one of the substances produced when red blood cells die.

One function of the liver is to break down bilirubin. Ordinarily, the liver gets rid of bilirubin by adding it to bile. Bile then flows into the small intestine, where it helps digest fatty food. Quite often, however, the liver of a newborn baby is unable to handle this aspect of its job. For this reason, jaundice in new babies is a common occurrence. In most jaundiced newborns the condition clears up by itself in a few days. As a baby feeds, its body begins to make all the chemicals needed for proper digestion, including bile.

Doctors sometimes expose jaundiced newborns to bright blue or white fluorescent light. Since bilirubin is light sensitive, exposure to light reduces its concentration in tissues. If jaundice does not return after the excess concentrations of bilirubin have been removed from the tissues, physicians are reasonably sure that the liver is functioning normally. Jaundice can also occur in children and adults who suffer from diseased livers.

A Rainbow of Colors

Bruises, or ecchymoses, are discolorations and tender areas of skin or mucous membranes. Unlike cyanosis, a bruise is caused by an injury that creates leakage of blood from vessels into tissues. Bruises can result from bumping into something that ruptures capillaries under the skin. A black eye is a type of bruise. Also, bruising can appear after energetic exercise in which jarring causes small rips in blood vessels.

When blood seeps into the surrounding tissue, it causes discoloration. Immediately after an injury a bruise appears red because of blood trapped under the skin. A day or two later red blood cells in the bruise begin to decompose, and the injured area darkens to blue or purple. By day six cells have broken down to the point that the area appears green. On day eight or nine the bruise takes on a brown or yellow color. Eventually, after the body completely breaks down the noncirculating blood and absorbs it, the bruised area returns to normal color. Because one can predict the rate at which a bruise will discolor, the shade can be used to predict time of injury. Generally, bruises are harmless and heal quickly. They usually fade within fourteen days. If they do not, a more serious problem may exist.

Sometimes a bruise hardens and increases in size. This can happen when blood becomes trapped in tissue and cannot be absorbed. Such a condition is called a hematoma. If a hematoma persists, it can be treated by draining the injured area.

Bruises differ in size and cause. Those that form deep purple patches as a result of a disease are called purpura. Sometimes the tendency to form purpura runs in families. Women are more susceptible than men. Often, purpura form because of the lack of some nutrient in the diet that provides one of the essential clotting factors. Older people tend to form senile purpura on the backs of their hands and forearms.

Petechiae are unusual, very small bruises. They may appear as tiny red dots in the skin. Petechiae are often seen in the whites of the eyes after periods of violent vomiting. If a person has died from unknown causes, presence of petechiae in the skin of the face and neck can indicate strangulation.

Some medications increase the tendency to bruise. A nonsteroidal anti-inflammatory like aspirin can make bruising worse because it thins the blood. Prescription blood

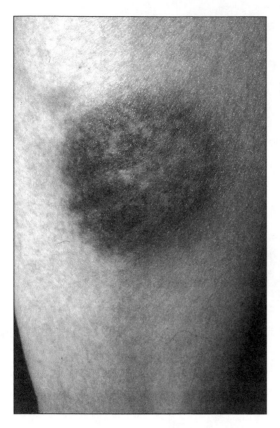

Bruises are caused by leaking or torn blood vessels.

thinners like warfarin have the same effect.

A Little Light-Headed

Enlarged veins that function poorly can cause other problems. Anytime blood cannot get back to the heart, either because of distended veins or for other reasons, fainting can occur. Fainting is a temporary loss of consciousness caused by a reduction of blood flow to the brain.

At one time it was not considered out of the ordinary for a lady to "swoon" or faint at least once a day. When women wore tight corsets, the garments interfered with blood returning to the heart. Therefore, there was not enough blood to pump to the brain, and fainting resulted.

Today, fainting, also known as syncope, is rarely caused by tight corsets. However, there are plenty of other types of problems that interfere with return of blood through the venous system and can result in fainting. Also, arrhythmia can cause a drop in heart rate due to inefficient pumping, leading to fainting. Tumors or blood clots in the heart can also interfere with its ability to pump.

A special type of fainting, orthostatic fainting, is triggered by standing up quickly. A rapid move from lying down to standing can cause as much as 20 percent, about one quart, of blood to be retained in the lower body. Consequently, blood volume returning to the heart is low, and the amount pumped to the brain is inadequate to sustain consciousness. A related form, parade ground fainting, happens when a person remains still for a long time. Without the movement of leg muscles to push blood back to the heart, blood pools in the lower extremities. Military guards and members of choirs may experience parade ground fainting during long ceremonies.

Fainting can also happen when nerves send signals to the circulatory system. For example, pain in the abdomen can stimulate the vagus nerve. It in turn sends a message to the heart to slow down, causing fainting. This type of fainting is called vasomotor or vasovagal syncope. It can also be stimulated by extreme fear and may be the mechanism that causes some animals, like the American opossum, to "play dead" when pursued. Some people experience vasovagal fainting at the sight of blood.

When the body gets very hot, venodilation or expansion of veins occurs. Veins near the skin dilate to radiate heat. As blood collects in the venous system, the heart cannot supply enough blood to the brain to keep the person conscious. Venodilation due to heat is nicknamed "Jacuzzi syncope" because it often happens when people lounge in very hot water. This condition may be very dangerous because the loss of consciousness in water can lead to drowning.

One can also faint after a period of straining. Straining increases pressure in the abdomen and chest, preventing return of venous blood from these areas. Straining can be caused by weight lifting, breath-holding, prolonged exhalations, coughing, bowel movements, or urination. Fainting due to straining is more common in older people.

Blood on Demand

The circulatory system does not work alone. It requires the help and support of all of the body systems. It is especially influenced by signals from the nervous system. One duty of the nervous system is to make sure that the areas of the body that need blood the most are supplied.

In the morning, when a person eats breakfast, the nervous system prompts the body to increase the amount of blood traveling to the digestive system. Later in the day, when that same person goes for a jog, the nervous system sends the signals necessary to increase the amount of blood traveling to muscles. That evening, when the person is reading, the nervous system requests that extra blood be sent to the brain.

The brain is part of the central nervous system, and the brain stem is in charge of many body activities, like blood allocation, that do not require conscious awareness. The brain stem uses two nerve pathways to regulate blood flow: the sympathetic and parasympathetic nervous systems. The sympathetic nerves are also called the "fight or flight" system. Activation of the sympathetic nerves prepares the body for action. Sympathetic nerves affect the circulatory system by increasing the heart rate as much as five times over its normal resting rate. At the same time, arterioles and arteries tighten to help propel blood quickly around the body. As a result, blood pressure increases dramatically.

The human body cannot remain in this heightened, ready-for-action state for long periods. When the danger or problem has passed, the parasympathetic nerves take over. The parasympathetic nerves are also called the "housekeeping system" because they maintain the body in a normal, nonemergency state. To prevent damage to the circulatory system, parasympathetic nerves lower heart rate and relax muscles in the arteries, causing blood pressure to return to normal.

Blood Pressure and Hypertension

"Blood pressure" refers to the force that blood exerts on vessels. The amount of pressure in a blood vessel is due to two factors: the strength with which the heart pumps blood and the resistance that vessel offers to the flow of blood. If blood pressure is not normal in all parts of the body, symptoms of the imbalance are quick to appear. Fainting, for example, is always preceded by a drop in blood pressure.

Blood pressure varies throughout life, and changes during each day depending on a person's level of activity. Adults have much higher blood pressure than babies and children. Activity, like exercise, raises blood pressure. And blood pressure is higher in the mornings than it is later in the day.

It is estimated that more than 50 million Americans suffer from high blood pressure, which is also called hypertension.

The sympathetic nervous system increases blood pressure during stressful situations by speeding the heart rate and constricting the arteries. However, it is difficult to detect when an individual's blood pressure is abnormally high. Contrary to popular opinion, high blood pressure does not cause redness in the face or headaches. However, it does damage the vessels and heart by increasing their wear and tear. High blood pressure increases a person's risk of stroke, heart failure, heart attack, and kidney disease.

Blood pressure is determined with a device called a sphygmomanometer. There are many variations of this device that range from the traditional cuff and stethoscope versions to the newer electronic devices. The original sphygmomanometers measured pressure by the distance in millimeters it pushed a column of mercury (Hg) up a tube. Even though many newer electronic blood pressure devices do not contain mercury, millimeters and Hg are still used as units of measurement.

Traditional blood pressure devices are known for their exceptional accuracy and can still be seen in many medical facilities. These units consist of a cuff with an inflatable bladder, a rubber bulb, and a measuring device. To check one's blood pressure, the inflatable cuff is wrapped around the upper arm, where the arm's major artery is only inches from the heart. Using the rubber bulb, the cuff is inflated to a point where the artery of the arm is compressed and blood flow through it is momentarily halted. Air from the cuff is slowly released until pressure falls to the point where blood flow through the artery resumes. This point, called the systolic pressure, is accompanied by a faint "rushing" noise that can be heard through a stethoscope placed on the arm over the artery. This first blood pressure reading is the maximum pressure of the blood. Air from the cuff continues to be released until pressure in the artery exceeds the pressure in the cuff. At this point no sound is heard at all through the

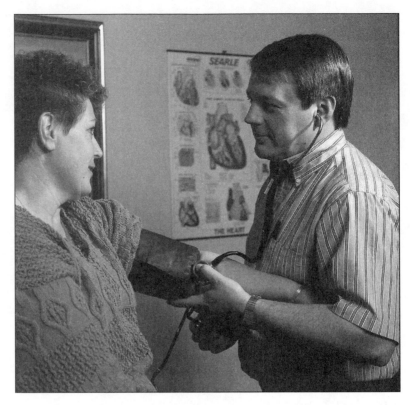

A physician determines a patient's blood pressure using a sphygmomanometer.

stethoscope. This second reading, the diastolic pressure, is the minimum value of the blood pressure.

Each time the heart contracts, it forces blood into an artery. After the period of heart contraction, the systole, the heart relaxes. During the relaxation period, the diastole, the pressure in the arteries decreases. A person's blood pressure is expressed as a ratio of the systolic pressure in the artery of the arm that was tested to the diastolic pressure in that artery. For example, normal blood pressure in an adult is 120/80 mm Hg (read as one twenty over eighty millimeters of mercury). Blood pressure is considered to be elevated if systolic is 140 mm Hg or more, and diastolic 90 mm Hg or higher.

Hypertension can be caused by several factors. Blood pressure can become elevated when the heart pumps with more force than usual, putting out more fluid than normal. Also, if arteries lose their flexibility, they become unable to stretch and recoil with each beat. Therefore, the blood from each heartbeat must force its way through a smaller space, increasing pressure. The addition of fluid to the circulatory system can also increase blood pressure. People whose kidneys are unable to remove enough water and waste components from the body tend to have high blood pressure.

About 90 percent of the time, doctors are unable to determine the cause of patients' high blood pressure. In these people the condition is described as primary hypertension. Hypertension in the remaining 10 percent may be due to kidney disease, hormones, stress, and obesity. In such cases it is described as secondary hypertension. No matter what the cause, it cannot be cured. Treatment to lower pressure includes weight loss in obese people, moderate exercise, changes in lifestyle that lower stress, and changes in diet for those with diseased arteries. Several medications help lower pressure by mechanisms such as reducing heart rate, reducing blood volume, or dilating arteries.

Pumping Up the Heart Muscle

Engaging in moderate exercise is one technique that can be used to help lower blood pressure. In general, exercise provides a lot of benefits to the cardiovascular system.

Exercise uses muscles, and active muscles need more oxygen than those at rest. To meet the demands of muscles during exercise, the heart beats faster, increasing the rate at which blood travels through the lungs and back to the tissues. High volumes of blood traveling through the heart during exercise cause it to work harder, improving its pumping ability.

The heart is a muscle, and like any muscle it increases in size with exercise. An athlete in peak condition may have a heart more than 40 percent larger than a person of comparable size who has not been training intensively. The strong, muscular heart of an athlete is able to pump a lot of blood through the body with relatively little effort. As a result, oxygen is delivered to cells faster.

Exercise helps the cardiovascular system.

Red, White, and Blue

Like bruises that are slow to heal, misconceptions about the heart, blood, and blood vessels still exist. One common error is the idea that blood traveling through veins is blue in color. Another is that the heart is shaped like the standard valentine. Some people still associate the heart with feelings, although in fact emotions are controlled by the nervous system, not the circulatory system. Nevertheless, changes in heart rate are one of the most noticeable responses to an emotional issue.

The appearance of the veins can change over time. Big veins that carry blood from the feet and hands to the heart can become weak and stretched. When these veins form lumpy cord-like knots under the skin, they are known as varicose veins. If the veins simply break and create blue lines in the skin, they are called spider veins. Neither condition is dangerous, but their appearance is considered unsightly by some.

Vessels just under the surface of the skin that are ruptured by injury can release blood and form bruises. Fresh bruises are red, but their color changes over time as the red blood cells within them decompose. The age of a bruise can be determined by its color.

If blood flow to the brain is reduced for any reason, a person might feel faint. Anything that restricts the flow of blood from the extremities back to the heart can cause fainting. Likewise, an activity that causes blood to pool in the legs, like prolonged standing, can also result in a dizzy spell. Fainting is always associated with lowered blood pressure. Blood pressure can be temporarily elevated by exercise or by a response of the sympathetic nervous system, the "fight or flight" branch of the brain. After a period of readiness, the parasympathetic system restores blood pressure to its normal level, thus protecting the integrity of the heart and blood vessels.

When the Circulatory System Malfunctions

4

Since the circulatory system provides nutrients and oxygen to all parts of the body, diseases and disorders that affect it can hamper the performance of the human body in many ways. Malfunctions may interfere with the work of the heart, blood, blood vessels, or any of the organs they support. In 1994 the American Heart Association reported that one in every five people suffered from some form of heart or blood vessel disease. Indeed, cardiovascular disease is the leading cause of death in the United States, claiming more than one million lives a year.

Some of the disorders related to the circulatory system can be prevented or postponed through healthy living strategies such as a balanced diet and a good exercise program. Other types of circulatory dysfunction are caused by factors that are hard to control such as genetics, infectious agents, and environmental triggers.

The Vital Pump

Disorders of the human heart can impact the heart muscle, tissues that line or cover the heart, the valves, or the coronary arteries. Tissues that cover the interior and exterior surfaces of the heart are subject to inflammation by chronic disease or by invasion of microorganisms such as bacteria, viruses, or fungi. In a condition called pericarditis, the two membranes covering the heart, the pericardia, thicken and rub against each other. As a result the heart has less space in which to contract and expand. Swelling of the pericardia can cause fever, chills, fatigue, and a sharp pain in the center and left part of the chest. Eventually, fluid may collect between the two membranes, making breathing painful.

Treatment for pericarditis depends on the causative agent. Bacterial infections respond to antibiotics while viral infections do not. In both cases, anti-inflammatory drugs help reduce swelling. Pericarditis can become a very serious condition if the volume of fluid in the pericardium increases to a point that it interferes with heart function, or if prolonged inflammation leads to the formation of scar tissue.

The innermost lining of the heart and its valves can also become inflamed. This condition, called endocarditis, occurs when microbes like bacteria gather in colonies within the lining. The presence of bacteria causes the body to send in white blood cells to destroy them. People whose valves have been damaged by prior heart disease are most susceptible to endocarditis. Some early warning signals of this disease are weakness, slight fever, aching joints, and heart murmurs. Treatment involves use of antibiotics for several weeks. When the disorder is caught early, chances of recovery are good.

Cardiac problems are not all due to infection or disease. Some people are born with defective hearts. Most such abnormalities are in the septum, which divides the heart into two halves. The job of the septum is to keep blood traveling through the right side of the heart separated from blood flowing through the left side. However, if there is an opening in the septum, an unwanted mix of oxygenated and deoxygenated blood can occur.

One of the most common defects of the septum is often reported in premature babies: The foramen ovale, the small window that allows fetal blood from the right atrium to enter the left atrium, bypassing the undeveloped lungs, may not be closed at birth. In some newborns the heart is able to compensate and send enough blood to the lungs to oxygenate the body. Other infants need surgery to close the hole in the septum.

The Gatekeepers

Disease processes can alter the diameter of one or more valves of the heart. Both narrowing and widening of heart valves puts additional strain on the heart. If a valve stiffens and narrows, the heart has to work very hard to push blood through it. On the other hand, if a valve widens, it cannot close properly once blood has passed through, so blood flows back through the valve in the wrong direction. Such leaky valves put strain on the heart by forcing it to pump not only the normal volume but also the additional blood that flowed back through the defective valve.

Doctors may first suspect that a heart valve is damaged if they hear a whooshing sound called a heart murmur while listening to a patient's heartbeat through a stethoscope. Some people who have heart murmurs suffer no ill effects. However, most patients with a damaged valve experience fatigue, breathlessness, and chest pain.

If test results show that heart valves are greatly damaged, they may have to be surgically repaired or replaced. Injured valves can be replaced with synthetic (man-made) valves, or

Severely damaged heart valves can be replaced with synthetic ones.

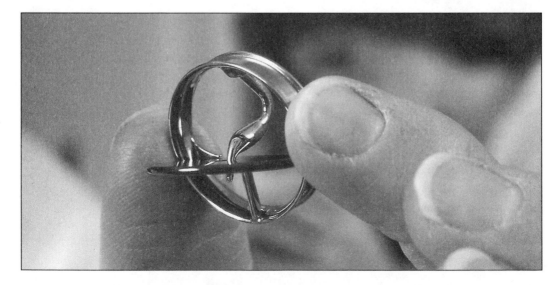

with valves from a pig's heart. Synthetic valves usually last longer than those from a pig, but they are more prone to cause blood clots. Both types of valves are susceptible to bacterial infections. Everyone who has heart valve replacement surgery must take antibiotics before future surgical or dental procedures to forestall bacterial infection that can further damage the valves.

In the early to mid–1900s, rheumatic fever was the primary culprit for damaged values. Rheumatic fever most often occurred in children between the ages of five and fifteen years who had previously suffered from strep throat. The streptococcal bacteria that infected their throats also invaded other organs such as the heart, joints, brain, and skin. The resulting condition, rheumatic fever, caused severe damage to heart valves in some children. The advent of antibiotic treatment for strep throat has made rheumatic fever a rare occurrence.

Heart valve disease can sometimes progress to an even more severe condition called congestive heart failure. In congestive heart failure, the heart does not fail completely, but functions very inefficiently. Blood flow out of the heart slows because the heart is unable to work at full capacity. Because the heart cannot pump enough blood to meet the body's needs, many organs of the body do not receive enough oxygen to function correctly, causing fatigue and weakness. The inability of the heart to pump plenty of blood through the body can also cause blood to back up into the lungs, making breathing difficult. Blood may even back up in the veins, causing fluid to leak into the surrounding tissue. This retention of fluid in body tissues causes swelling in areas of the lower body such as the legs and feet.

Symptoms of congestive heart failure can be mild, moderate, or severe. Doctors can often diagnose the problem after a physical exam. A chest X ray and other imaging techniques can help confirm the diagnosis. Treatment depends on the severity of the condition. Mild or moderate cases of congestive heart failure can usually be treated with medication and changes in lifestyle. This includes drugs that improve the flow of blood through the heart and diuretics to help the body eliminate excess water. Simple choices such as avoiding salty food and elevating the legs when sitting can also diminish symptoms. If the heart muscle itself is not damaged, surgical valve replacement can sometimes correct severe cases of congestive heart failure. If the heart is damaged, a heart transplant may be the only option.

The Heart's Own Supply

The coronary arteries, which supply the heart muscle with blood, are also subject to disease. Coronary artery disease (CAD) occurs when the coronary arteries narrow, reducing the flow of blood. Hereditary factors such as diabetes and high blood pressure, as well as environmental agents like fatty diets and smoking tobacco products, can harm the walls of arteries. Usually, however, narrowing is due to the buildup of deposits of plaque within the vessels. Damage within a vessel attracts platelets, which are soon joined by fat, cholesterol, and cell wastes. Eventually white blood cells form a fibrous cap over the accumulated material, creating the deposit known as plaque. Over time, further deposits enlarge the plaque deposit until blood flow is hampered. The restriction of blood flow through a vessel due to buildup of plaque is called arteriosclerosis.

Narrowing of coronary arteries is a slow process that occurs over a period of several decades. It often begins during childhood, but gets rapidly worse after the mid-thirties. A healthy artery is like a length of clean pipe. In diseased arteries the pipe becomes clogged, causing the lumen to narrow.

In the early stages CAD often causes no symptoms. Most people who are diagnosed with CAD do not learn of the condition until the artery has narrowed to 30 percent of its original size. When oxygen-carrying blood cannot reach the heart, chest pain called angina may result. After a vessel is narrowed, any occurrence that puts an extra burden on the heart and increases its need for oxygen can result in pain.

Both physical exertion and stress can trigger angina. Angina is often described as heavy pressure, burning, or aching over the middle of the chest. Sometimes this pain can be relieved by rest and relaxation. However, if coronary arteries are narrowed to less than 20 percent of their original size, angina may be experienced even at rest.

Angina, along with shortness of breath, is a warning sign of serious heart problems. Both congestive heart failure and CAD can trigger a heart attack. Heart attack, or myocardial infarction, occurs when a portion of heart muscle dies. In the United States heart attacks are the most common cause of death.

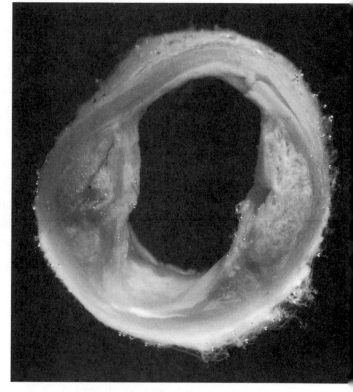

A view of a partially obstructed artery.

Once a heart attack has begun there is a three-hour window before the affected heart muscle actually dies. If the heart can regain normal function in that time frame, damage to heart muscle can be avoided. For that reason medical experts recommend that a person suffering severe angina for more than twenty minutes seek immediate medical attention.

Not everyone suffers from the same kind of angina during a heart attack. Some people experience little or no pain, while others report a sensation similar to indigestion. Many people feel a squeezing pressure on their chest that spreads up to the jaw or down the arm. Some suffer from sweating, dizziness, and shortness of breath. Recent research suggests that if a person thinks they are experiencing a heart attack, they should chew an aspirin while they wait for medical attention. It has been shown that aspirin thins the blood, helping it flow through the heart.

Many people who suffer from mild or moderate CAD are instructed to eat a healthy diet, get plenty of moderate exercise, and practice stress reduction techniques to prevent further accumulation of plaque in arteries. Medications that lower high blood pressure and those that help blood flow easily through vessels may be recommended.

Out of Rhythm

CAD can cause life-threatening abnormal heart rhythms called arrhythmias. Not all arrhythmias are dangerous. In fact, everyone's heart skips a beat or flutters at one time or another. However, dangerous arrhythmias occur when the heart cannot meet the demands placed on it.

Arrhythmias can cause the heart to beat too quickly or too slowly. Irregularity may appear in the atria, the receiving chambers of the heart, or the ventricles, the pumping chambers. The most serious type of arrhythmia is called ventricular tachycardia (V-Tac). In V-Tac, a ventricle receives fast and disorganized electrical impulses. Instead of contracting and pumping blood, the ventricle begins to quiver, a state called fibrillation. When a ventricle is quivering, it is not pumping blood. The pulse of a person in V-Tac may accelerate from two to four times that of normal heart rhythm. If fibrillation is not stopped, cardiac arrest can occur and the heart will quit beating.

Abnormal rhythms of the ventricles cause 90 percent of all cardiac arrests. If a person in cardiac arrest is not revived within minutes through CPR (cardiopulmonary resuscitation) or by application of an electric shock to the heart, sudden cardiac death occurs. Many times, blockage of two or more coronary arteries leads to sudden cardiac death. Even if a person is revived after cardiac arrest, portions of their heart may have been damaged. A victim's prognosis depends on what area, and how much, of the heart was damaged.

Plaque in Other Places

The coronary arteries are not the only blood vessels that can become clogged and damaged. Any artery in the body can develop plaque. The carotid arteries, which carry blood from the heart to the brain, are especially prone to plaque deposits. As plaque builds up, the arteries narrow and the flow of blood through them slows. This condition, called carotid artery disease, can progress to the point that all blood flow to the brain is blocked. Without a supply of blood, brain tissue dies.

Arteries in the legs can also develop arteriosclerosis. Deposits of plaque inside the arteries leading to muscles in the legs cause a condition called peripheral artery disease. People who smoke, as well as those with high blood pressure, diabetes, and high cholesterol, are more prone to peripheral artery disease than the general population. Some symptoms of this condition include a cramping, heavy feeling in the legs below the areas of blockage. The pain is especially noticeable during exercise, when oxygen demands are elevated. Doctors can diagnose the condition using imaging techniques. If left untreated, blood flow in the affected leg can become completely blocked. Surgical correction may be needed to prevent death of tissue. If tissue death occurs, amputation of the limb is the only option.

Arteriosclerosis can also lead to aneurysms, weakened segments of blood vessels. Aneurysms usually form on areas of an artery that are damaged by plaque formation. The buildup of plaque inside an artery prevents that region of the artery from receiving nourishment and causes it to weaken. As blood passes over the weakened area, the damaged blood vessel stretches, ballooning outward. Blood may even pool in the resulting widened section of artery. Some people are born with weakened segments in blood vessels, but most who receive this diagnosis have developed the abnormality over time.

The symptoms associated with an aneurysm depend on its location in the body. These weakened areas can form anywhere, and some, although not all, are life-threatening. The most common site of an aneurysm is in the largest artery of the body, the aorta. Other aneurysm-prone locations include the

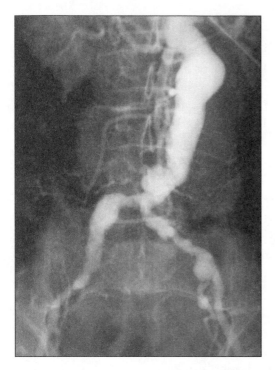

An X ray shows an abdominal aneurysm.

small arteries of the brain and arteries in the legs.

If any aneurysm ruptures, the person experiences severe and sudden pain. Bleeding, shock, coma, and possibly death may follow. When a doctor suspects an aneurysm, tests are performed to determine its size and exact location. Surgery is necessary in cases where the bulge is large and rupture is likely. People with aneurysms are instructed to keep their blood pressure within normal limits because high blood pressure can place additional strain on the already weakened section of artery. They may also be warned to avoid tobacco products and fatty foods.

The Road Is Blocked

Pieces of plaque sometimes break loose from the walls of the arteries. If they travel through the bloodstream to another location, they can become lodged in a smaller artery, blocking it completely. Any tissue that does not receive blood flow can die.

A blood clot can also block a vessel. A clot that forms inside of a damaged artery is called a thrombus. Thrombi may break free and clog other vessels, but they may also remain attached to their point of origin. An embolus, on the other hand, is a blood clot that moves through the circulatory system. Blood flow to any organ or limb can be blocked by an embolus, but some of the most common sites are legs, kidneys, and the brain. Strokes result when blood flow to or within the brain is obstructed. When circulation to the arms or legs is blocked, these extremities cease to have a pulse. They can become pale, weak, numb, and cold. If the blockage is not corrected, tissue death can occur, a condition that requires amputation of the affected limb. Not all emboli result in tissue death because vessels running parallel to the damaged artery help supplement the blood supply to the deprived area.

Blood clots can develop in veins just as they do in arteries. Clots that form in the large veins of the legs are called deep vein thrombi (DVT). They result when blood stagnates in the lower veins of the legs. Blood is ordinarily moved from the feet back toward the heart by the squeezing action of leg muscles, very much as toothpaste is squeezed out of a tube. People who cannot walk or otherwise use their leg muscles because they are restricted to bed rest are often victims of this deep vein thrombosis. If blood clots in the veins of the legs break free, they travel directly to the heart, which pumps them into the lungs. Because the lungs are a maze of tiny vessels, the clot usually gets stuck there.

Blood clots in the lungs trigger a condition called pulmonary embolism, a serious situation that can cause chest pain, shortness of breath, rapid heart rate, fainting, and even death. Doctors can diagnose DVTs with imaging techniques that show the size of the clot and the speed of blood flowing around it. Patients suffering from pulmonary embolism are often given medication that thins the blood so that it can travel around the clot more freely. Sometimes surgical removal of the clot is required.

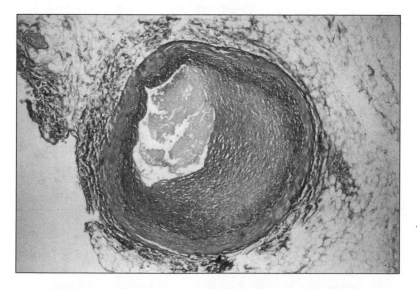

A blood clot, as seen in the center of this artery, can cause much damage.

Although they rarely cause serious problems, veins near the surfaces of the legs are also subject to disease. Phlebitis is inflammation of the veins near the skin. It sometimes follows injury to a vein. Symptoms include heat, redness, swelling, and pain around the infected area. Doctors can usually diagnose the condition after a physical exam. Patients are often instructed to use warm compresses and anti-inflammatory medications to relieve pain and reduce swelling. Strong elastic stockings provide compression on the veins and help keep blood moving through them.

The hands and feet can also suffer from circulatory problems. In a disorder called Raynaud's disease, small capillaries in the skin contract or narrow, causing blood flow to the hands and feet to be temporarily restricted. When blood flow is stopped, fingers and toes turn white. As all of the oxygen in these tissues is used up, the color changes to blue. Restriction of blood flow makes the fingers and toes feel numb or prickly. Stimuli such as cold air and stress often trigger the symptoms of Raynaud's disease. Blood flow can be restored by gradually warming the affected areas.

A Battle with Blood Cells

The trillions of blood cells in the human body are also subject to malfunction. When the body does not produce enough red blood cells, anemia results. The job of red blood cells is to transport oxygen to all parts of the body. Levels of hemoglobin, the protein in red blood cells that carries oxygen, is reduced in people who suffer from anemia. Doctors can diagnose the condition with a blood test that checks levels of hemoglobin.

There are several types of anemia, and they vary in cause and severity. Mild forms of anemia may produce no symptoms. On the other hand, severe forms can cause a person to feel tired, short of breath, and light-headed. People with anemia sometimes have headaches, and their heart rates are accelerated because their bodies are working hard to make up for the lack of red blood cells.

In the most common type of anemia in the United States, iron deficiency anemia, there is not enough iron in the body to make the amount of red blood cells needed to take care of the patient's oxygen transport needs. Iron is a vital component of hemoglobin. Iron deficiency anemia can be caused by several conditions, including insufficient iron in the diet, loss of blood through injury or chronic disease, cancer, and medications that cause internal bleeding. Occasionally, women who are menstruating heavily lose enough blood to cause anemia.

From Past Generations

Some people are born with a genetically transmitted condition called sickle-cell anemia, a disease caused by an abnormal type of hemoglobin. This strange hemoglobin forms unusual, long chains. Red blood cells containing this form of hemoglobin become distorted, taking on a crescent or sickle shape, and tend to rupture easily. The rapid

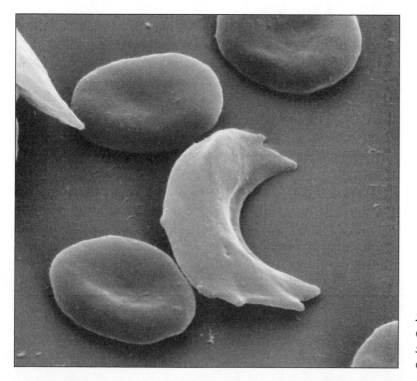

Sickle-cell anemia can cause crescent-shaped red blood cells.

and constant loss of red blood cells due to rupture stimulates the body to make new red blood cells. However, they are produced at such a fast rate that they enter the bloodstream in an immature and nonfunctional state.

Red blood cells that are distorted by sickle-cell anemia often get stuck in capillaries, blocking them and preventing the flow of blood to tissues. The type and severity of symptoms depends on where the blockage occurs. Victims often suffer pain in the arms, legs, and abdomen, as well as fever and ulceration of the skin. In severe cases convulsions and paralysis can result. There is no cure for the disease, and treatment only manages the symptoms.

Another genetically transmitted blood disease is hemophilia. Hemophiliacs, sometimes called "free bleeders," lack one or more of the blood factors needed for clotting. However, they have normal levels of platelets. There are several types of the disease, identified by which clotting factor is missing.

People suffering from hemophilia bleed excessively from even minor injuries. They also suffer from internal bleeding within muscles, joints, and body cavities. It is not unusual for someone with hemophilia to experience nosebleeds or blood in their urine. Hemophilia, which appears primarily in males but it is transmitted by females, is incurable. Symptoms are managed by transfusions of clotting factors, whole blood, or plasma.

Winding Up Circulatory Disorders

Even though the heart and its system of vessels are incredibly strong and durable, they are subject to damage and disease. Heart disease is the leading cause of death in the United States, claiming the lives of almost two thousand people each day.

Any condition that interferes with the flow of life-sustaining blood through the body can be critical. To keep the body functioning, its circulatory equipment, the heart and vessels, must move blood to and among all cells without interference.

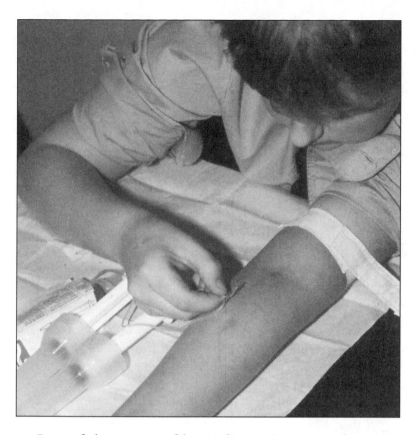

A boy suffering from hemophilia performs a blood transfusion on himself.

One of the causes of heart disease is arteriosclerosis, a narrowing of blood vessels. Arteriosclerosis, derived from the Greek words for "hard porridge," begins when the wall of a vessel is damaged. Ironically, the very platelets that accumulate at the site of damage lead to more problems because they attract and hold fatty droplets of cholesterol. The sticky deposit, called plaque, causes the artery to become brittle and weak. Sometimes the plaque triggers the growth of a blood clot that can travel through the circulatory machinery to wreak havoc elsewhere in the body.

If plaque blocks arteries in the heart itself, heart muscle is deprived of oxygen. Lack of oxygen produces pain known as angina, literally a "strangling in the chest." Angina may serve as a warning to stop activity and seek help. Sometimes, however, it precedes the death of heart cells, a condition known as heart attack.

Probing and Exploring the Circulatory System

5

The structures that deliver life-sustaining blood to every cell extend all through the body. Yet most of this blood transport system is concealed deep within tissues, not readily visible to the human eye. When problems occur with the heart, blood, and blood vessels doctors must often look deep inside to find the answers.

A physician is an investigator who gathers information in order to solve a problem. A cardiologist, a specialized type of physician, focuses on solving problems related to the heart and blood vessels. A cardiologist begins this search with an interview to find out what is bothering the patient. By compiling a medical history, a physician can learn about problems in the patient's past as well as those in the patient's family. The last, and most complex, stage of the investigation is the examination.

An examination may begin with simple procedures such as listening to the heart, checking the pulse, and measuring blood pressure. After that, there is a wide array of medical tests that can be performed to learn more about the patient's condition. Diagnostic tests such as blood analysis, electrical impulse monitoring, and imaging are commonly used.

Reading the Blood

One of the first tests performed to gather information about the circulatory system is a blood analysis. The condition and composition of blood can say a great deal about how well the circulatory system is functioning. Using a hypodermic needle, a small sample of blood can be drawn from a vein on the inside of the arm. The extracted blood is collected in sterile tubes and labeled with the patient's name, date, and names of tests to be performed.

Some blood tests are very general and provide doctors with information about the overall health of the patient. Blood work that measures hemoglobin levels, electrolytes (minerals dissolved in the blood), composition of gases dissolved in the blood, the ability of the blood to clot, and the number of red and white blood cells in the sample are some tests commonly ordered by physicians. Other blood tests are specific and give data from which the performance of different organs and tissues of the body can be analyzed in detail.

If doctors suspect that a patient may already have experienced a heart attack, they request a blood test that will indicate the presence of certain enzymes. These telltale cardiac enzymes are released into the bloodstream during a heart attack or when heart tissue is injured. One of the enzymes can be detected in as little as four hours after the heart attack, while the other one appears twelve to twenty-four hours later. If either of these enzymes is found in a sample of the patient's blood, the doctor's tentative diagnosis is confirmed and further evaluation of the heart is in order.

Checking Out the Beat

Blood work provides doctors with some answers, but more details may be needed before the source of a patient's problem can be identified. The heart's rhythm provides huge amounts of data about how well the heart is functioning. With every beat, the heart gives off electrical impulses that can be detected and recorded. The graphic record of the heart's rate and rhythm thus produced is called an electrocardiogram (ECG or EKG). To record the ECG, small patch electrodes are placed on the chest, arms, legs, and other parts of the body. Wires, or leads, connect these electrodes to a machine that collects and analyzes the information. The patient's beating heart causes the electrocardiograph to generate a series of wavy lines. By studying the size and shape of the waves as well as the time between the occurrences of each wave, doctors can spot an irregular heart rhythm and make other diagnostic determinations.

An ECG that is recorded when a patient is lying down is called a resting ECG. It indicates heart rhythm when a person is not stressed. In some diseases, however, the heart beats normally when the patient is at rest but goes into irregular rhythms with movement or exercise. To find out how the heart performs under stress, an ECG may be taken while the patient walks on a treadmill. This kind of ECG is sometimes referred to as an exercise stress test. Problems that are not evident during a resting ECG may become obvious when the heart is forced to beat faster, an event that requires the coronary arteries to provide additional oxygen to heart muscle. The occurrence of exercise-induced arrhythmia provides important clues about a heart disorder.

Some abnormal heart rhythms are more difficult to detect because they show up only occasionally. Doctors can

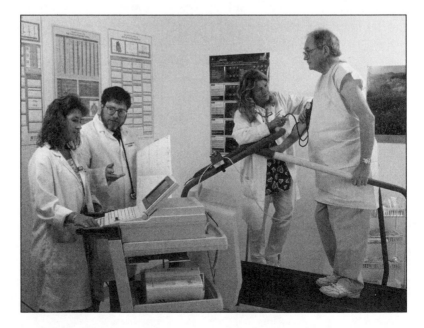

Stress tests are often performed on a patient in order to make the heart beat faster.

monitor the heartbeat for an extended period by having their patient wear a portable ECG. This device, called a Holter monitor, records heartbeats over a twenty-four-hour period. The battery-powered monitor is worn over the shoulder like a purse, and its electrodes are attached to the patient's chest. As the patient goes about daily activities, the heartbeats are recorded and analyzed. At the end of the twenty-four hours, the patient returns to the doctor's office to be disconnected and to turn in their data.

Seeing Inside

Imaging is another valuable tool that doctors can use to assess circulatory disorders. Imaging techniques, which range from simple two-dimensional X rays to complex Computerized Tomogrphy (CT) scans, provide pictures of structures inside the body. To evaluate the heart's shape and position in the chest, doctors may order chest films. These are images formed by X rays on a photographic screen and reproduced for viewing on a piece of sensitized film or on a computer. The X rays themselves are waves of energy that are absorbed by dense body structures but can easily travel through less dense tissue.

X rays that are not blocked by dense tissue move through the body and expose the photographic film, creating dark areas. However, X rays that are absorbed by bone, muscle, or other dense types of tissue create shadows on the film. Inspection of X rays helps visualize the position, size, and shape of the heart and blood vessels and very clearly shows conditions such as an enlarged heart or aorta.

CT scans are one of the most advanced types of imaging. A CT scanner transmits multiple X rays through the body at different angles. Special sensors in the scanner then measure the amount of radiation absorbed by different tissues in the area being studied. For a CT image of the heart, a computer program uses the differences in X ray absorption to form a three dimensional, cross-sectional view of the heart and blood vessels.

Looking with a Magnet

A magnetic resonance imaging system (MRI) uses magnets instead of X rays to provide a detailed look inside a patient's body. During an MRI session, the patient is surrounded by electromagnets that magnetize the body, causing all of the hydrogen atoms to line up in one direction. Hydrogen atoms are more abundant in some types of tissues than in others. Different types of tissues respond to magnets by absorbing, then reflecting, radio waves of different frequencies. The body is then exposed to a shower of radio waves. A computer collects and analyzes these re-emitted radio frequencies, creating a color picture on a computer screen. Areas of high hydrogen atom density can be clearly distinguished from less dense regions.

MRI images are often used to find out how much damage was done to heart muscle after a heart attack. They can also reveal conditions such as aneurysms, pericarditis, and cardiac lesions. Since pacemakers are influenced by the pull

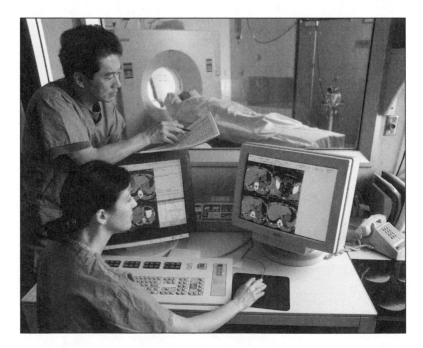

A patient is sent into an MRI scan. MRIs are used to find heart conditions.

of magnets, people in whom these devices have been implanted generally cannot have MRIs.

Looking at Sound Waves

An echocardiogram is a cardiac image produced by sound waves. Echocardiography, also called a cardiac echo, is one of the most widely used techniques in diagnosing heart problems because it is noninvasive, does not use X rays, and provides excellent results. It is especially useful in finding abnormalities of heart valves and other disorders in cardiac structure.

During the test a gel or oil is spread on the patient over the area to be studied. A pencil-like probe called a transducer is rubbed in the gel. As it moves, it gives off high frequency sound waves that travel into the body. These waves bounce off internal structures, such as the heart, creating a series of echoes. The returning echoes are electronically plotted on a computer monitor, creating a picture called an echocardiogram. Such a picture shows doctors the condition of the heart's walls, valves, and blood vessels. It is most frequently used to diagnose aneurysms and to help monitor diseases of the heart. To visualize structures in the back of the heart, the doctor may pass the transducer down the patient's esophagus.

Sometimes an echocardiogram is followed by a similar test, using Doppler ultrasonography. This test, which measures blood flow through major vessels, is performed very much like an echocardiogram; it is often used to evaluate the condition of the carotid arteries, large vessels that lead from the heart to the brain. By knowing volume and speed of blood flow, doctors can determine whether or not the carotid arteries are blocked by either blood clots or plaque. Echocardiogram and Doppler ultrasound techniques provide a lot of information about the heart and major arteries.

Radioactive Pictures

Pictures produced by imaging techniques can often be enhanced by the use of contrast media, materials that show up well on X-ray film. When a contrast medium travels through an organ, it helps outline that organ, making it easier to study. Before nuclear imaging scans are performed, the patient must walk on a treadmill to increase his or her heart rate and get blood moving rapidly through the body. Then the contrast medium, often radioactive thallium, is injected into a vein that carries blood to the heart. Next, a gamma camera, which detects radioactive energy, is focused on the patient's chest to track the movement of the radioactive substance as it travels through the bloodstream. Output from the camera helps doctors determine if heart muscle is adequately supplied with blood. It also enables them to see how well each chamber of the heart is functioning.

On the images produced by the gamma camera, damaged places show up as "cold spots," areas where radioactive substance could not penetrate. These places represent coronary arteries that are clogged or tissue that has been injured by a heart attack. This test also gives the doctor information on how the heart performs when it is stressed by exercise. After the first set of images is recorded, the patient is asked to rest for an hour or two. Then a second set of pictures is made, this time recording the heart at its normal resting rate.

Results of the thallium scan tell a valuable story. If both the resting and the exercise test readings are normal, the patient's coronary arteries are clear and in good working order. If the resting test is normal but the exercise test is not, one or more of the coronary arteries may be partially blocked, preventing the heart muscle from getting enough oxygen when it is stressed. If both the resting and the exercise thallium results are abnormal, there may be considerable blockage in the coronary arteries. If thallium does not move through all of the heart muscle, part of the heart is dead due to a previous heart attack.

Outlining with Chemical Dyes

If one or more imaging tests give abnormal results, a cardiologist may order an additional imaging check called a cardiac catheterization. This test, which is more invasive than simpler imaging techniques, allows doctors to get a picture of both the heart and the coronary arteries using X rays and a contrast dye that outlines the internal surfaces of the arteries.

During a cardiac catheterization, the patient reclines on a table that is positioned under X-ray equipment. A small incision is made in either the patient's arm or leg. A long, thin, flexible tube called a catheter is inserted through the incision into an artery or a vein, depending on which portion of the heart is being studied. A special X-ray technique called fluoroscopy helps the doctor guide the catheter through the blood vessels until it reaches the part of the heart to be studied. During fluoroscopy, a continuous stream of X rays is passed through the patient's body and projected onto a fluorescent screen to allow visualization of the beating of the heart muscle. Once the catheter tip

A doctor watches a screen that shows the interior of the blood vessel.

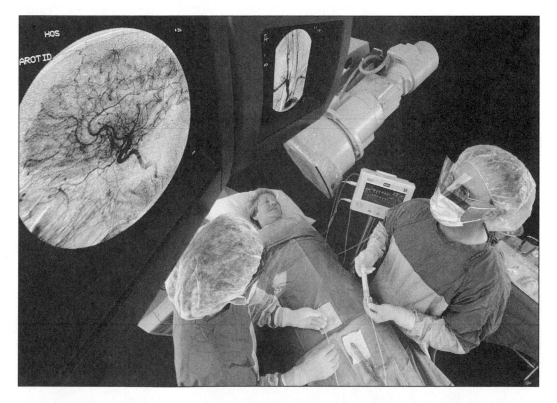

reaches its destination, the contrast dye is injected through the catheter and into the coronary arteries, producing an outline of the arteries on a video screen. Moving X rays are recorded as the dye filters into the coronary arteries. These moving pictures, called angiograms, clearly show any blockages in the arteries of the heart.

A cardiac catheterization, being invasive, is more dangerous than some other tests. During the process there is a possibility that the catheter will dislodge a blood clot or plaque deposit from the inside of a blood vessel. Depending on where the clot or plaque travels, such a mistake could cause a heart attack. A small percentage of the population are allergic to chemical dye used in the test and suffer unpleasant symptoms.

Chemical dyes are used to study other parts of the circulatory system, too. Venography images the veins of the legs and feet to help determine the presence of obstructions or to locate a suitable vein that can be used as a graft in coronary bypass surgery. During venography, contrast dye is injected into the large vein of the foot or leg. An X-ray machine above the leg records the path of the dye as the patient is asked to push his or her foot against a solid object to force blood into the veins. Blockages such as deep vein thrombi or other abnormalities can be seen on the moving image produced during the test. As with other tests using dyes, there is a slight risk that the patient will suffer an allergic reaction.

Pressing Plaque Against the Wall

When results of an angiogram show significant blockage in one or more coronary arteries of the heart, cardiologists may recommend a procedure to improve the flow of blood to the heart. Angioplasty is one way to open up narrowed coronary arteries.

The process begins the same way as a heart catheterization with a catheter being guided to the blocked artery. In angioplasty a second balloon-tipped catheter is passed through the first catheter. When the balloon tip reaches the blocked area of the artery, the balloon is inflated and deflat-

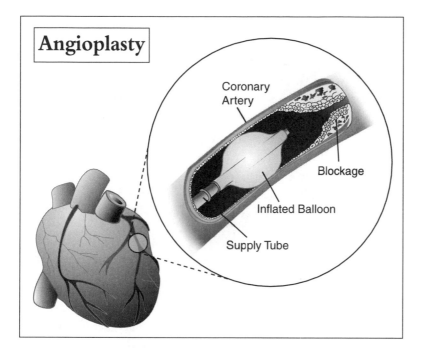

Angioplasty

Coronary
Artery

Blockage

Inflated Balloon

Supply Tube

ed several times. This action pushes the plaque against the inner artery wall, widening the lumen to allow for improved blood flow. Once the artery is opened the catheters are withdrawn one at a time, and the patient is restricted to bed rest for at least the next eight hours.

Statistics show that about one-third of patients who have angioplasty develop blockages in that same location within six months after the procedure. Consequently, surgeons sometimes used a stent in conjunction with angioplasty. The stent is a short piece of tubing made from mesh-like metal and fitted on the balloon tip of the catheter. Once the catheter reaches the location of the blockage, the balloon is inflated, causing the stent to expand in diameter. When the balloon is deflated and withdrawn, the stent remains in place, serving as a permanent scaffolding for the now widened artery. By supporting the artery, the stent helps to prevent closure of the vessel, which occurs soon after some balloon angioplasty procedures. Only 15 to 20 percent of patients with stents have further blockages at that location within six months after the procedure.

Not all people with blockages in coronary arteries are candidates for simple angioplasty. If the blockage in a coronary artery is so large that the artery is almost closed, a balloon may not fit in the narrowed space. In this case, a cardiologist may use a fiber-optic probe attached to the tip of the catheter to cut away part of the plaque. After some of the obstruction has been cleared, balloon angioplasty can be done. Another option for patients with severe blockage is a procedure called an atherectomy. A tiny drill-like device attached to the end of a catheter is advanced to the blockage in the artery. Once in position, it pulverizes plaque to a very small size that can pass through the circulatory system.

A Trip Around the Blockage

Sometimes blockages in coronary arteries are so numerous that coronary bypass surgery is the only option. For such a procedure, the cardiologist must send the patient to a doctor who specializes in vascular surgery. The purpose of coronary bypass surgery is to reroute blood around clogged arteries, improving the supply of blood and oxygen to the heart.

During the surgery a healthy blood vessel is removed from another part of the body such as the leg or chest. Removal of this vessel for use as graft material does not deprive the body of blood because most areas of the body are supported by several vessels. One end of the severed vessel is sewn below the coronary artery blockage, and the other end is attached above it in the same way. As a result, blood can flow around the obstruction very much like cars might take a detour to avoid fallen trees on a highway. Often, more than one coronary vessel is blocked, so the surgeon must perform multiple bypass grafts.

Before bypass surgery the patient is given general anesthesia. The surgeon cuts through the skin, muscle, and breastbone to get to the heart. Since this procedure interferes with the heart's ability to pump blood, the patient's heart is stopped and blood is routed through a heart-lung machine, which pumps blood through the body and keeps it

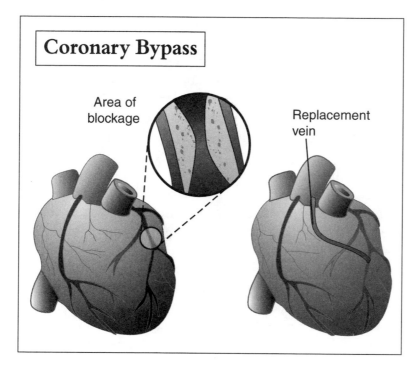

Coronary Bypass

Area of blockage

Replacement vein

oxygenated, in effect taking over the jobs of both sides of the heart. Once the grafts are sewn into position, the heart is gently massaged so it will return to its normal pattern of contracting and relaxing. As soon as the heart begins beating again, the heart-lung machine is disconnected, the patient's breastbone is wired together, and the skin and muscle are stitched into place. Patients usually remain in the hospital four or five days after surgery. Medical personnel will suggest some healthy lifestyle changes for the patient to consider. Common suggestions include a low fat diet, a regular exercise program, and abstinence from tobacco products. These changes in lifestyle help prevent the grafts from getting clogged with plaque.

Getting a New Heart

Heart disease can sometimes damage parts of the heart, causing some of its tissue to die. If damage to the heart is severe, the patient may need a heart transplant. The transplanted heart can either be artificial or living tissue. In the case of a living heart, the tissues of a transplant recipient must be matched to the tissues of a donor.

Recipients are selected based on blood type, body weight, severity of illness, and geographical location. Location is a very important consideration because a human heart cannot be disconnected from a donor's circulation for more than about four hours without losing the ability to work properly in an intended recipient.

Most heart donors are healthy people who have been fatally injured. As a result, their brains are dead, but the organs in their chest cavity are still alive. When a donor heart becomes available, a heart transplant team removes the organ, puts it in a cold preservative solution, and transports it to the hospital where the recipient is waiting.

Meantime, the recipient is taken to surgery where his or her chest is opened and blood is diverted by a heart–lung machine. The recipient's diseased heart is carefully detached from its blood vessels, which are reattached to the new organ. Once the operation is complete, the new heart is massaged so it begins to beat again. When a heartbeat is established, the heart–lung machine is disconnected and the patient's chest is closed.

The greatest danger of heart transplant is rejection. Normally a person's body would perceive the tissue of a new heart or other organs as "foreign" and would launch an attack against it. To suppress this natural reaction, the recipient of a heart is put on antirejection medications for the remainder of his or her life.

The wait for a heart transplant can extend for months or even years. Patients whose hearts are so badly damaged that they cannot survive a long wait are sometimes given a temporary artificial heart. This option buys more time for the patient while a donor heart is sought. The use of artificial hearts began in the 1980s. One of the best known artificial hearts is called the Jarvik-7, named for the inventor.

The Jarvik-7 works like a natural heart. It has two pumping chambers that serve as ventricles. Made of plastic, aluminum, and polyester, it connects to a system of com-

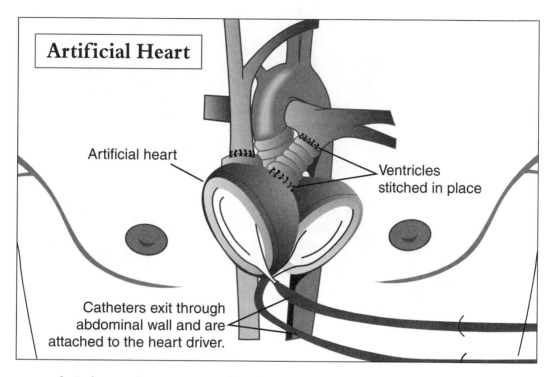

Artificial Heart

Artificial heart

Ventricles
stitched in place

Catheters exit through
abdominal wall and are
attached to the heart driver.

pressed air hoses that pass into the chest cavity. The heart
and hoses are powered by an external energy system. Risk
of infection with the Jarvik-7 and all artificial hearts is high.
Therefore, they are used only temporarily. Researchers predict that one day artificial hearts will be perfected so that
they can permanently replace a damaged heart.

Keeping the Pace

The heart depends on its natural pacemaker, a cluster of
nerve tissue in the right atrium, to keep the heart beating
in a regular rhythm. If this electrical system breaks down,
the results can be fatal. Many heart attacks are caused by a
failure in the cardiac electrical system.

An electronic pacemaker can sometimes take over the job
of a flawed natural one. These battery-operated devices are
surgically implanted under the skin of the chest. Wires
extend from the pacemaker to the heart. When the heart's
rhythm slows below a set rate, the pacemaker kicks in to
restore the normal beat.

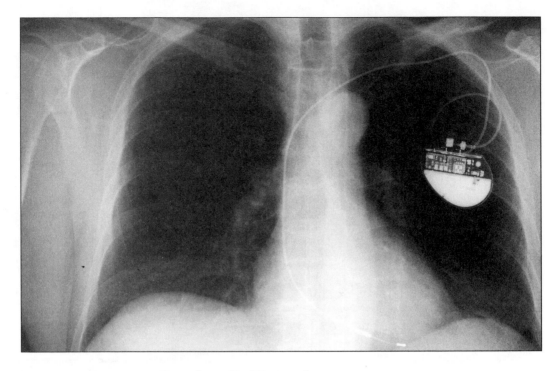

An X ray shows a pacemaker in a patient's heart.

Sewing It Together

Early physicians knew that blood was vital to life, but they did not understand its role in the body. Until the early nineteenth century, bloodletting to remove bad "humors" was a treatment for many diseases. Often it was performed by applying bloodsucking worms called leeches to patients' skin. Another technique was cupping, which could be done wet or dry. Wet cupping involved making shallow cuts in the skin. Then a piece of paper or fabric was burned inside a cup, which was quickly inverted over the area of cuts. Burning in the cup created a vacuum, which drew blood to the surface. In dry cupping the skin was not cut because the pull of the vacuum created by the fire under the overturned receptacle was considered to be strong enough to remove vile "humors" from the blood. Leeches were also used to remove "excess" blood from sick patients.

Today, doctors know that removing large quantities of blood makes a patient sicker instead of better. Physicians now understand how the circulatory system works to deliver blood to the body. When there is a problem with the system, they perform tests. Laboratory analysis, imaging techniques, and electric conductivity studies are routinely used to establish a diagnosis. Once the problem is known, patients have many options for correcting it. Some diseases can be reversed simply by changes in lifestyle. However, others require medication or more drastic action such as angioplasty or open-heart surgery.

GLOSSARY

antibody: Special protein made by the body as a defense against a foreign material.

antigen: Anything that causes the body to launch an immune response through the production of antibodies.

aorta: Largest artery in the body.

arrhythmia: Abnormal heart rhythm.

artery: Vessel carrying blood away from the heart.

atrium: Upper chamber of the heart that receives blood.

blood type: Blood category based on presence or absence of certain antigens on red blood cells.

capillary: Small vessel that connects an artery to a vein.

cardiac: Relating to the heart.

coronary arteries: Vessels supplying blood to the heart.

deoxygenated: Without oxygen.

diastole: Resting stage of a heartbeat.

diffusion: Movement of molecules from an area of high concentration to one of lesser concentration.

hemoglobin: A pigment of red blood cells responsible for the transport of oxygen.

leukocyte: White blood cell.

lumen: The space within a tubular structure, such as that space within a blood vessel.

pericardium: Membrane surrounding the heart; an inflammation of this membrane is called pericarditis.

plaque: Fatty deposit on blood vessel walls in atherosclerosis.

plasma: Liquid part of blood.

systole: Contraction phase of a heartbeat.

vein: Vessel carrying blood away from the heart.

ventricle: Lower chamber of the heart that pumps blood.

FOR FURTHER READING

Books

Elizabeth Fong, *Body Structures and Functions.* St. Louis, MO: Times Mirror/Mosby College, 1987. Comprehensive but concise description of basic human anatomy and physiology.

Alma Guinness, *ABC's of the Human Body.* Pleasantville, NY: Reader's Digest Association, 1987. Guinness describes physiological processes in reader-friendly terms.

The Handy Science Answer Book. Canton, MI: Visible Ink Press, 1997. Answers some frequently asked questions about the human body and other areas of science.

How in the World? Pleasantville, NY: Reader's Digest Association, 1990. Provides short explanations of how body structures work.

David E. Larson, *Mayo Clinic Family Health Book.* New York: William Morrow, 1996. Explains common injuries, diseases, and disorders.

Susan McKeever, *The Dorling Kindersley Science Encyclopedia.* New York: Dorling Kindersley, 1994. Summarizes many scientific principles, including those that govern the human body.

Medicine. Eyewitness Books. London: Dorling Kindersley, 1995. Discusses many common medical problems.

World Book Medical Encyclopedia. Chicago: World Book, 1995. Broad collection of diseases and disorders.

Websites

American Academy of Family Physicians (AAFP) (www.familydoctor.org). The AAFP promotes and maintains high-quality standards for family doctors providing comprehensive health care to the public. Information on the causes and symptoms of arrhythmia is available.

The American Heart Association (www.americanheart.org). The mission of the American Heart Association is to reduce disability and death from cardiovascular diseases and stroke. This site gives information on the causes of atherosclerosis and ways to prevent it.

Dr. Koop (www.drkoop.com). The former U.S. surgeon general's site provides information on all areas of health and disease. It includes descriptions on the effects and treatments of bleeding.

HowStuffWorks (www.howstuffworks.com). HowStuffWorks is a media company that is internationally recognized as the leading provider of information on how things work. The site offers an excellent description of heart function.

LifeClinic (www.lifeclinic.com). The LifeClinic website was developed by medical experts to provide an in-depth resource for information about prevalent, long-term health conditions such as high blood pressure, diabetes, or asthma.

Internet Sources

Mayo Clinic, "Blood's Color: Is it Blue or Red?" February 10, 2002. www.mayoclinic.com.

American Association of Blood Banks, "Blood Supply and Utilization Data Statement," April 20, 2001. www.aabb.org.

WORKS CONSULTED

Books

Robert Berkow, *The Merck Manual of Medical Information.* New York: Pocket Books, 1997. Written for the layman, this guide presents basic science information and explains diseases and their treatments.

Body Systems: Anatomy and Physiology. New York: Macmillan, 1993. Reviews body systems and their functions.

Jennifer Cochrane, *The Illustrated History of Medicine.* London: Tiger Books International, 1996. Beautiful book that features many of the pioneers in medicine.

Charlotte Dienhart, *Basic Human Anatomy and Physiology.* Philadelphia: W.B. Saunders, 1979. Classic presentation of anatomy, physiology, and diseases.

William C. Goldberg, *Clinical Physiology Made Ridiculously Simple.* Miami, FL: Med Masters, 1995. Light-hearted, simplistic explanations of body system structures.

John Hole Jr., *Essentials of Human Anatomy and Physiology.* Dubuque, IA: Wm. C. Brown, 1992. Comprehensive review of human anatomy and physiology.

The Incredible Machine. Washington, DC: National Geographic Society, 1986. Fascinating and awe-inspiring description of inner workings of the human body.

Ann Kramer, *The Human Body, The World Book Encyclopedia of Science.* Chicago: World Book, 1987. Explains how the human body works.

Anthony L. Komaroff, *Harvard Medical School Family Health Guide.* New York: Simon & Schuster, 1999. Layman's book of health.

Stanley Loeb, *The Illustrated Guide to Diagnostic Tests.* Springhouse, PA: Springhouse, 1994. Describes how and why many diagnostic tests are used.

Roberto Margotta, *The History of Medicine.* New York: Reed International Books, 1996. Examines how medicine has changed over the centuries.

Elaine Marieb, *Human Anatomy and Physiology.* Redwood City, CA: Benjamin/Cummings, 1995. Excellent text on anatomy and physiology presented in informal writing style.

Websites

National Library of Medicine (www.nlm.nih.gov). Provides information on all medical topics, as well as links to other medical resources.

American Academy of Family Physicians (AAFP) (www.familydoctor.org). The AAFP promotes and maintains high-quality standards for family doctors providing comprehensive health care to the public. Information on the causes and symptoms of arrhythmia is available.

The American Heart Association (www.americanheart.org). The mission of the American Heart Association is to reduce disability and death from cardiovascular diseases and stroke. Lists and describes types of congenital heart defects.

The American Association of Blood Banks (www.aabb.org). A resource center on blood collection. Describes how blood banks collect and assess blood.

The Texas Medical Center (www.tmc.edu). A medical school that provides information for the public. Explains a variety of topics on heart conditions.

Texan Arrhythmia Institute (www.txai.org). The Texas Arrhythmia Institute is a nonprofit organization with a simple but vital mission: the prevention and treatment of cardiac arrhythmias. Explains mechanisms that cause fainting.

ENT Associates (www.entassociates.com). The Ear, Nose and Throat Associates in Corpus Christi, Texas, provide information for the public. Describes the care and prevention of nosebleeds.

Medtronic (www.medtronic.com). Medtronic is the world leader in medical technology providing lifelong solutions for people with chronic disease. Explains symptoms and diagnostic processes for evaluating syncope, as well as information on modern pacemakers.

INDEX

PICTURE CREDITS

ABOUT THE AUTHORS

Both Pam Walker and Elaine Wood have degrees in biology and education from colleges in Georgia. They have taught science in grades seven through twelve since the mid-1980s.

Ms. Walker and Ms. Wood are coauthors of more than a dozen science teacher resource activity books and two science textbooks.